Advanced Processing and Manufacturing Technologies for Structural and Multifunctional Materials V

Advanced Processing and Manufacturing Technologies for Structural and Multifunctional Materials V

A Collection of Papers Presented at the 35th International Conference on Advanced Ceramics and Composites
January 23–28, 2011
Daytona Beach, Florida

Edited by
Tatsuki Ohji
Mrityunjay Singh

Volume Editors
Sujanto Widjaja
Dileep Singh

A John Wiley & Sons, Inc., Publication

Published by John Wiley & Sons, Inc., Hoboken, New Jersey.
Published simultaneously in Canada.

For general information on our other products and services or for technical support, please contact our Customer Care Department within the United States at (800) 762-2974, outside the United States at (317) 572-3993 or fax (317) 572-4002.

Wiley also publishes its books in a variety of electronic formats. Some content that appears in print may not be available in electronic formats. For more information about Wiley products, visit our web site at www.wiley.com.

Library of Congress Cataloging-in-Publication Data is available.

ISBN 978-1-118-05993-7

oBook ISBN: 978-1-118-09537-9
ePDF ISBN: 978-1-118-17270-4

ISSN: 0196-6219

Contents

Preface

The Fifrth International Symposium on Advanced Processing and Manufacturing Technologies for Structural and Multifunctional Materials and Systems (APMT) was held during the 35th International Conference on Advanced Ceramics and Composites, in Daytona Beach, FL, January 23-28, 2011. The aim of this international symposium was to discuss global advances in the research and development of advanced processing and manufacturing technologies for a wide variety of non-oxide and oxide based structural ceramics, particulate and fiber reinforced composites, and multifunctional materials. This year's symposium also honored Professor Katsutoshi Komeya, Yokohama National University, Japan, recognizing his outstanding contributions to science and technology of advanced structural and multifunctional ceramics and his tireless efforts in promoting their wide scale industrial applications. A total of 66 papers, including invited talks, oral presentations, and posters, were presented from more than 10 countries (USA, Japan, Germany, China, Korea, Turkey, Ireland, Slovakia, Slovenia, Serbia, Canada, India and Israel). The speakers represented universities, industry, and research laboratories.

This issue contains 19 invited and contributed papers, all peer reviewed according to The American Ceramic Society review process. The latest developments in processing and manufacturing technologies are covered, including green manufacturing, smart processing, advanced composite manufacturing, rapid processing, joining, machining, and net shape forming technologies. These papers discuss the most important aspects necessary for understanding and further development of processing and manufacturing of ceramic materials and systems.

The editors wish to extend their gratitude and appreciation to all the authors for their cooperation and contributions, to all the participants and session chairs for their time and efforts, and to all the reviewers for their valuable comments and suggestions. Financial support from the Engineering Ceramic Division and The American Ceramic Society is gratefully acknowledged. Thanks are due to the staff of the meetings and publication departments of The American Ceramic Society for their invaluable assistance.

We hope that this issue will serve as a useful reference for the researchers and technologists working in the field of interested in processing and manufacturing of ceramic materials and systems.

TATSUKI OHJI, *Nagoya, Japan*
MRITYUNJAY SINGH, *Cleveland, USA*

Introduction

This CESP issue represents papers that were submitted and approved for the proceedings of the 35th International Conference on Advanced Ceramics and Composites (ICACC), held January 23–28, 2011 in Daytona Beach, Florida. ICACC is the most prominent international meeting in the area of advanced structural, functional, and nanoscopic ceramics, composites, and other emerging ceramic materials and technologies. This prestigious conference has been organized by The American Ceramic Society's (ACerS) Engineering Ceramics Division (ECD) since 1977.

The conference was organized into the following symposia and focused sessions:

Symposium 1	Mechanical Behavior and Performance of Ceramics and Composites
Symposium 2	Advanced Ceramic Coatings for Structural, Environmental, and Functional Applications
Symposium 3	8th International Symposium on Solid Oxide Fuel Cells (SOFC): Materials, Science, and Technology
Symposium 4	Armor Ceramics
Symposium 5	Next Generation Bioceramics
Symposium 6	International Symposium on Ceramics for Electric Energy Generation, Storage, and Distribution
Symposium 7	5th International Symposium on Nanostructured Materials and Nanocomposites: Development and Applications
Symposium 8	5th International Symposium on Advanced Processing & Manufacturing Technologies (APMT) for Structural & Multifunctional Materials and Systems

Symposium 9	Porous Ceramics: Novel Developments and Applications
Symposium 10	Thermal Management Materials and Technologies
Symposium 11	Advanced Sensor Technology, Developments and Applications
Symposium 12	Materials for Extreme Environments: Ultrahigh Temperature Ceramics (UHTCs) and Nanolaminated Ternary Carbides and Nitrides (MAX Phases)
Symposium 13	Advanced Ceramics and Composites for Nuclear and Fusion Applications
Symposium 14	Advanced Materials and Technologies for Rechargeable Batteries
Focused Session 1	Geopolymers and other Inorganic Polymers
Focused Session 2	Computational Design, Modeling, Simulation and Characterization of Ceramics and Composites
Special Session	Pacific Rim Engineering Ceramics Summit

The conference proceedings are published into 9 issues of the 2011 Ceramic Engineering & Science Proceedings (CESP); Volume 32, Issues 2–10, 2011 as outlined below:

- Mechanical Properties and Performance of Engineering Ceramics and Composites VI, CESP Volume 32, Issue 2 (includes papers from Symposium 1)
- Advanced Ceramic Coatings and Materials for Extreme Environments, Volume 32, Issue 3 (includes papers from Symposia 2 and 12)
- Advances in Solid Oxide Fuel Cells VI, CESP Volume 32, Issue 4 (includes papers from Symposium 3)
- Advances in Ceramic Armor VII, CESP Volume 32, Issue 5 (includes papers from Symposium 4)
- Advances in Bioceramics and Porous Ceramics IV, CESP Volume 32, Issue 6 (includes papers from Symposia 5 and 9)
- Nanostructured Materials and Nanotechnology V, CESP Volume 32, Issue 7 (includes papers from Symposium 7)
- Advanced Processing and Manufacturing Technologies for Structural and Multifunctional Materials V, CESP Volume 32, Issue 8 (includes papers from Symposium 8)
- Ceramic Materials for Energy Applications, CESP Volume 32, Issue 9 (includes papers from Symposia 6, 13, and 14)
- Developments in Strategic Materials and Computational Design II, CESP Volume 32, Issue 10 (includes papers from Symposium 10 and 11 and from Focused Sessions 1, and 2)

The organization of the Daytona Beach meeting and the publication of these proceedings were possible thanks to the professional staff of ACerS and the tireless dedication of many ECD members. We would especially like to express our sincere thanks to the symposia organizers, session chairs, presenters and conference attendees, for their efforts and enthusiastic participation in the vibrant and cutting-edge conference.

ACerS and the ECD invite you to attend the 36th International Conference on Advanced Ceramics and Composites (http://www.ceramics.org/daytona2012) January 22–27, 2012 in Daytona Beach, Florida.

SUJANTO WIDJAJA AND DILEEP SINGH
Volume Editors
June 2011

SEEDS INNOVATION AND BEARING APPLICATIONS OF SILICON NITRIDE CERAMICS

Katsutoshi Komeya
Yokohama National University, Yokohama, Kanagawa, Japan

ABSTRACT

Silicon nitride (Si_3N_4) ceramics based on the Y_2O_3 and Al_2O_3 addition have been recognized as the most attractive materials for the wear resistant applications as bearing balls. Then it was found that the addition of TiO_2 to the Si_3N_4-Y_2O_3-Al_2O_3 originated the densification promotion and the improvement of the cyclic fatigue life. In the sintered bodies, TiN grain changed from TiO_2 was located in the grain boundary. Most recently the author's group has developed new Si_3N_4 ceramics designed with nano-sized TiN dispersion, which shows lower damage for mating material in the bearing system. The development of the highly reliable bearing balls using the newly innovated seeds was provided as the collaborative research with the industries. The key processing factor is to design and control nano-dispersion of TiN in the sintered bodies. In order to develop nano-sized TiN-dispersed Si_3N_4, a mechanochemical dry-mixing technique has been adapted to produce Si_3N_4-TiO_2 composite powder, which was mixed with other sintering aids by wet-ball milling in the following step. The sintering behavior of the powder compacts was examined in details, and sintered bodies were evaluated on the targeted items required for reliable bearing materials. Consequently, we have achieved the targeted values; performance of Class 1 in ISO26602, low aggressiveness to a mating metals (SUJ2), longer lifetime over 10^7 cycle under the load of 5.9 GPa by thrust type rolling fatigue test.

INTRODUCTION

As is well known, Si_3N_4 has been widely used as one of typical engineering ceramics, because its strong covalent bond structures make features of excellent characteristics, such as heat resistance, corrosion resistance, high hardness, low thermal expansion coefficient, high thermal conductivity and so on. However, it has no melting point at around 1900°C under ordinary atmospheric pressure, high vapor pressure and very low diffusion coefficient. These intrinsic properties of Si_3N_4 suggest that this material has poor sinterability to full density. Therefore, special sintering techniques such as reaction sintering, hot-pressing, gas pressure sintering, HIPping and SPS (spark plasma sintering) have been developed for densification of this material.

Historically, actual development of Si_3N_4 ceramics started from the reaction

1

sintering of Si_3N_4 by Collins et al.[1] in 1955. Various shapes of sintered bodies with high porosity were fabricated. Deeley et al.[2] created dense sintered bodies by hot-pressing of the Si_3N_4-MgO system for the first time in 1961. Komeya, et al.[3-5] discovered rare-earth-oxides, especially Y_2O_3, as sintering aids for densification in 1969-73.

Si_3N_4 ceramics with high strength and fracture toughness have been developed using the composition of α-Si_3N_4-Y_2O_3-Al_2O_3 and simultaneous innovation such as discovering SiAlONs[6, 7], the development of fine, pure, and highly sinterable Si_3N_4 powder[8], the invention of a gas pressure sintering technique[9], the advancement of the science and technology for microstructure control etc. It has been confirmed that the SiAlONs are expressed by the rational formulas of $Si_{6-z}Al_2O_zN_{8-z}$ (z = from 0-4.2)[10] for β-sialon, and $M_xSi_{12-(m+n)}Al_{m+n}O_nN_{16-n}$ (M: Ca, Y, Yb ...., x = m/ν, ν: electric charge of M, n: number of substitutional oxygen) for α-sialon[11, 12]. Therefore, the sintered Si_3N_4-Y_2O_3-Al_2O_3 composition is considered to be composed of SiAlONs grains and SiAlON glass phases. Since 1980s, Si_3N_4 ceramics have been applied to automobile components such as glow plugs[13], hot chambers[14], and turbocharger rotors[15]. Around the same time, cutting tools and bearing components were also developed[16-18]. Furthermore, Si_3N_4 ceramics fabricated by TiO_2 added to the Si_3N_4-Y_2O_3-Al_2O_3 have been widely used for bearing applications. Therefore, the author and his co-workers have precisely studied the sintering behavior and microstructure control by TiO_2, and most recently we have developed new Si_3N_4 ceramics designed with nano-sized TiN dispersion, which show lower damage for mating material in the bearing system, as introduced later.

BACKGROUND IN THE DEVELOPMENT OF SILICON NITRIDE BEARING

Toshiba has been involved in materials innovation and applications development in Si_3N_4 ceramics for many years, during which time wear materials development and their applications, especially ceramic bearings, have been a most successful effort, which continue to expand in area and volume. There are two types of ceramic bearings, hybrid bearing and all ceramic bearing. The first application was achieved as a ceramic bearing for a machine tool spindle, followed by various other applications such as small bearings for computer hard-disk drives (HDDs), antimagnetic bearings for semi-conductor production, and bearings for main engines in the space shuttle. As a similar application, Toshiba and Cummins, Inc. developed a Si_3N_4 injector link for a diesel engine based on significant wear reduction during operation[19].

A more precise description of the first bearing applications of Si_3N_4 ceramics is as follows. Joint research between Toshiba and Koyo Seiko began in the early 1980s. Prior to that, Koyo Seiko had completed a feasibility study of various kinds of ceramics such as Si_3N_4, SiC, Al_2O_3 and ZrO_2 by using thrust type tests for evaluating the fatigue life of rolling bearings. In these tests, a typical high-strength Si_3N_4 with a sintering aid of $5wt\%Y_2O_3-2wt\%Al_2O_3$, hot-pressed at 1700 to 1800°C under 30 MPa in 0.1 MPa N_2, which was supplied by Toshiba, retained a rolling life equivalent to or better than that of steel (SUJ2) bearings, which is shown in Fig. 1[16]. Significantly, no defects such as voids or impurities were observed at the damage starting points. Damaged parts had the same fatigue spalls observed with conventional steel bearings, encouraging mechanical engineers to pioneer new applications. The practical application was achieved in a bearing for a machine tool spindle, since the material had the important advantage indicating lower temperature rise for higher rotation speed in spite of its high cost (Fig.2[20]).

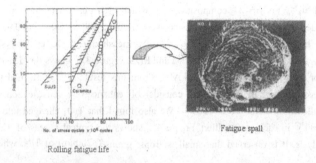

Rolling fatigue life   Fatigue spall

Fig. 1 Rolling fatigue life and fatigue spall of hot-pressed $92wt\%Si_3N_4-5wt\%Y_2O_3-2wt\%Al_2O_3$ by thrust type rolling fatigue test.

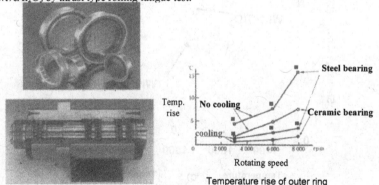

Temperature rise of outer ring

Fig. 2 Hybrid bearings for machine tool spindle and the temperature rise of the outer ring during operation.

TiN DISPERSED Si3N4 CERAMICS AS BEARING MATERIAKS [5,6,18]

In 1980, Komatsu et al[21] found that TiO_2 addition to the Si_3N_4-Y_2O_3-Al_2O_3 system promoted the densification at lower temperatures. Toshiba has made efforts to develop the new material as conventional bearing materials. After then, it was confirmed that the durability over the indispensable repetition load required as sliding parts was improved remarkably. Thus, TiO_2 doped Si_3N_4 ceramics have been expanded to the various bearing applications, which is shown in the previous paragraph.

Recently Yokohama National University focused on investigating the role of TiO_2 addition on the sintering of the Si_3N_4-Y_2O_3-Al_2O_3 and effect of the existence of TiN derived from TiO_2 on the mechanical properties. Experimental results are described as follows.

Starting compositions of 0-5wt%TiO_2 and 0-5wt%AlN were prepared for addition to typical Si_3N_4-Y_2O_3-Al_2O_3 compositions, in which high purity and sub-μm sizes of raw powders were used. Powder compacts were fired at 1300 to 1900°C in 0.9MPa N_2. Sintered bodies were evaluated by density measurement, XRD analysis and microstructural analysis using SEM, TEM and EDS. Figure 3[22] shows the densification curves for with and without TiO_2 and AlN. From this result, it is confirmed that the simultaneous addition of TiO_2 and AlN considerably enhanced the densification, with an almost full density achieved at 1600°C. We also found that TiO_2 changed into TiN at about 1000°C in each composition. Figure 4[23] shows the typical results of TEM and EDS analysis. It is observed that small isotropic grains (~200nm) of TiN, which was

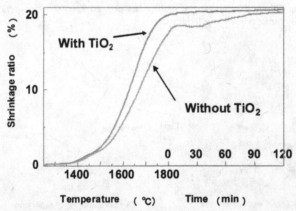

Fig. 3 Shrinkage curves for the Si_3N_4-Y_2O_3-Al_2O_3 with and without TiO_2 by the in-situ measurement using Laser dilatometer.

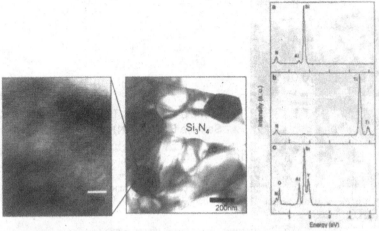

Fig. 4 TEM and EDS analysis of the sintered body of the Si_3N_4-Y_2O_3-Al_2O_3-AlN with TiO_2.

identified to be TiN by XRD profiles, is located in the grain boundary and many edge dislocations were observed in a high-resolution micrograph, suggesting that the TiN grain had been subjected to high stress during firing. Since the particle size of raw TiO_2 powder is about 170 nm, the size of TiN seems to reflect that of the raw TiO_2 powder.

Tatami et al.[24] have made it clear that improvement in the endurance by the TiN particles, as shown in the R-curve of Fig. 5, is because crack progress was controlled by improvement of the stress intensity factor in a minute crack region due to TiN. The contact damage testing method was adapted for tribological evaluation of the sintered specimens.

Fatigue test has been performed using cyclic loading test method. A tungsten carbide ball 1.98 mm in diameter was used as the loading indenter. In the experiment, ring-like cracks, called cone cracks, were formed on the surface of the Si_3N_4 samples by the loading. The result is shown in Fig. 6[24]. From the relationship between the number of loading cycles and the bending strength of the specimen after cyclic indentation under conditions of P=2500 N and f (frequency) = 10 Hz, we determined that the bending strength of the specimens without TiN particles declined as the number of cycles was increased, while that of TiN particle-dispersed Si_3N_4 specimens held the same value as before indentation. This means that the dispersion of the TiN particles prevents crack propagation under cyclic indentation. The limiting of crack propagation probably results from compressive residual stress at the grain boundary.

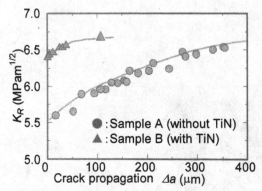

Fig. 5 R-curve behavior of sintered Si_3N_4 ceramics with and without TiN. Initial notch of 60% of the height of the test bar (35x3x1.5) was introduced (tip radius: 10μm).

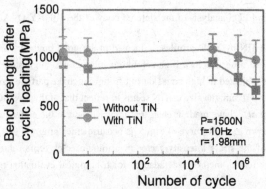

Fig. 6 Bending strength after cyclic loading of the sintered bodies of the Si_3N_4-Y_2O_3-Al_2O_3 with and without TiN: load; 1500N, frequency; 10Hz and indenter; WC ball.

Hybrid bearings composed of Si_3N_4 ceramic balls and metal rings are more popular than all-ceramic bearings because of their cost. However, it is possible that hard TiN particles damage the mating metals in a manner analogous to the wear map concept[25]. We have examined wear rates for the sintered bodies fabricated by different particle sizes of TiO_2 powders, which are 20, 170, 180, 540 and no TiO_2 addition. Figure 7 indicates the result with the largest size of TiO_2, 540 nm shows the largest wear rate for mating metal (SUJ2). However, the wear rate for 20 nm TiO_2 is the same as 170nm TiO_2, which seems to be based on the agglomerates of TiO_2 nanoparticles. In order to solve this problem, we have focused the development of the dispersion

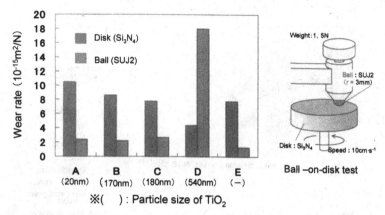

Fig. 7 Effect of TiO$_2$ particle size on wear resistance measured by ball-on-disk test.

technologies for TiO$_2$ nanoparticles.

DEVELOPMENT OF HIGHER RELIABILITY OF NANO TiN DISPERSED SILICON NITRIDE CERAMICS

TiN nanoparticles can be formed from TiO$_2$ nanoparticles because the size of TiN should be almost the same as that of TiO$_2$. However, it is difficult to realize TiN nanoparticle-dispersed Si$_3$N$_4$ ceramics even if TiO$_2$ nanoparticles are completely dispersed in the slurry[26]. This difficulty might be because of the reagglomeration of TiO$_2$ nanoparticles during the drying process. So, we have examined various different kinds of powder mixing techniques. Consequently, we found that formation of the powder composite composed of nano-TiO$_2$ bonded Si$_3$N$_4$ particle is the most advantageous approach to develop TiN-nanoparticle-dispersed Si$_3$N$_4$ ceramics and lower the aggressiveness to the mating metals in a wear test. Then, we have started the new joint work with Toshiba Materials Co. and JTKT Corp. under the NEDO contract: "Development of high performance TiN nano-particle dispersed silicon nitride ceramics for rolling bearings". Results of the project are shown below.

To fabricate TiN-nanoparticle-dispersed Si$_3$N$_4$ ceramics, 20 nm TiO$_2$ powder to the same Si$_3$N$_4$-Y$_2$O$_3$-Al$_2$O$_3$-AlN composition was used as the raw materials. First, TiO$_2$ nanoparticles were dispersed in ethanol according to our previous study[26]. The Si$_3$N$_4$ powder was then mixed into the TiO$_2$ slurry by ball milling, followed by the elimination of the ethanol. The pre-mixed powder, Si$_3$N$_4$ and TiO$_2$, was mechanically treated using a powder composer (Nobilta NOB-130, Hosokawa Micron Co., Japan), which is shown in

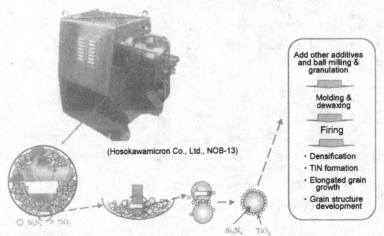

Fig. 8 Fabrication process to fabricate nano TiN dispersed Si_3N_4 ceramics using mechano-chemical dry mixing technique and equipment (Nobilta NOB-130, Hosokawa Micron Co., Japan). Nano-particle of TiO_2 is strongly bonded on the surface of Si_3N_4 by large shear stress derived from high speed rotation.

Fig. 9 SEM images of powder mixture (a) before and (b) after a powder composite process.

Fig. 8 to prepare composite particles. After the powder composite process, the other sintering aids were added and mixed by ball milling in ethanol. After removing the ethanol, 4 wt% paraffin and 2 wt% dioctyl phthalate (DOP) were added as a binder and lubricant, respectively. For reference, another powder mixture of the same composition with 20 nm TiO_2 particles and without TiO_2 was prepared by conventional ball milling.

Figure 9 shows SEM images of the powder mixtures before and after the mechanical treatment. The TiO_2 nanoparticles formed aggregates before the mechanical treatment (Fig. 9 (a)). On the other hand, as shown in Fig. 9 (b), no TiO_2 nanoparticle aggregates were found in the powder mixture after the mechanical treatment, i.e., the TiO_2 nanoparticles were well dispersed. Figure 10 presents TEM images of the powder

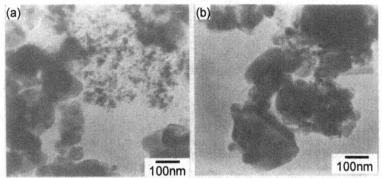

Fig. 10 TEM images of powder mixture (a) before and (b) after a powder composite process.

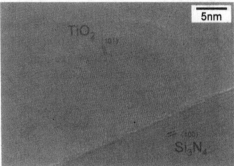

Fig. 11 High-resolution TEM image of composite particle prepared by a powder composite process.

mixtures before and after the mechanical treatment. Although the TiO_2 nanoparticles were dispersed in ethanol by wet mixing, they reagglomerated as a result of mixing with Si_3N_4 powder and/or drying. As shown in Fig. 10 (b), the mechanical treatment resulted in the uniform dispersion of the TiO_2 nanoparticles, thus suggesting that they might be strongly attached to the Si_3N_4 particles. A high-resolution TEM (HRTEM) image of a composite particle is shown in Fig. 11[27]. It can be seen that a TiO_2 nanoparticle is directly bonded onto a submicron Si_3N_4 particle. At the atomic scale, the interface between the TiO_2 and Si_3N_4 was flat. Such a direct-bonded interface should be stronger than the interface of physically adsorbed particles. In addition, neck growth occurred between the TiO_2 and Si_3N_4 particles, similar to the initial stage of sintering, in spite of the mechanical treatment at ambient temperature. This phenomenon should result from the reaction between the TiO_2 and SiO_2 and/or Si_3N_4.

After then, the mixed powders prepared by ball milling the composite powder and

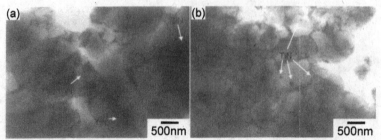

Fig. 12 TEM images of sintered Si_3N_4 ceramics: (a) 20nm TiO_2 and wet ball milling, and (b) 20nm TiO_2 and mechanochemical treatment.

the other sintering aids, Y_2O_3, Al_2O_3 and AlN, were sieved using a nylon sieve and then molded into ϕ 15 × 7 mm pellets by uniaxial pressing at 50 MPa followed by cold isostatic pressing at 200 MPa. After binder burnout, the green bodies were fired at 1800 °C in 0.9 MPa N_2 for 2 h. The sintered bodies were hot isostatically pressed at 1700 °C for 1 h under 100 MPa N_2. The density by Archimedes method and the phase present identified by X-ray diffraction were examined. The microstructure was observed by SEM and TEM-EDS. The Si_3N_4 ceramics were machined using a grinding machine and then polished using diamond slurry. The wear property was estimated by a ball-on-disk test, which consisted of a polished Si_3N_4 disk and a steel ball bearing (SUJ-2). The radius of the SUJ-2 ball was 3 mm. The rotation speed and radius were 10 cm s^{-1} and 3 mm, respectively. The weight was 5 N, and the running distance was 250 m. β-SiAlON and TiN were also identified as the main phases of the products in the sample, in addition to TiO_2 and AlN. Figure 12 shows TEM images of the TiN-dispersed Si_3N_4 ceramics. As shown in Fig. 12 (a), the size of the TiN particles in the Si_3N_4 ceramics fabricated using just wet mixing was 300–500 nm. On the other hand, in the case of using composite particles prepared by premixing and mechanical treatment, 20–100nm TiN nanoparticles were found in Si_3N_4 grains and in the grain boundary (Fig. 12(b)). Thus, it was shown that TiN nanoparticle-dispersed Si_3N_4 ceramics were fabricated using composite powder prepared by premixing followed by mechanical treatment.

Wear resistivity was evaluated from the wear volume of the TiN nanoparticle doped Si_3N_4 disk and SUJ2 balls after the ball-on-disk tests. As listed in Table 1, the wear volume of the Si_3N_4 disk was independent of the Si_3N_4 ceramics. On the other hand, the wear volume of the SUJ2 ball depended on the mating Si_3N_4 ceramics, i.e., the wear volume of the ball worn by sample fabricated using the composite powder, Si_3N_4 and TiO_2, was not only less than the half of the sample by ball milling method, but also

Table 1 Wear volumes of Si₃N₄ disks fabricated by different mixing process for 20nm TiO₂ and SUJ2 ball after ball-on-disk test.

Sample	Mixing process	Wear volume / $10^{-12}m^3$	
		Si_3N_4 disk	SUJ2 ball
TiO_2:20nm	Powder composite process	9.5	1.5
TiO_2:20nm	Conventional wet ball milling	9.8	3.9
No TiO_2	Conventional wet ball milling	10.0	1.6

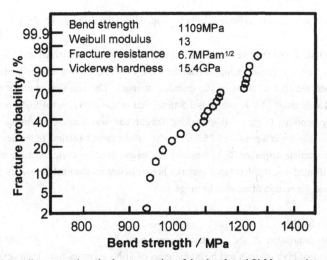

Fig. 13 Bending strength and other properties of the developed Si₃N₄ ceramics.

the same as that worn by sample without TiN, thus suggesting that TiN nanoparticle dispersion should cause less damage to the mating metals[28].

The developed Si₃N₄ ceramics need high mechanical properties for bearing application. ISO 26602 published in 2009 provides a classification defining the physical and mechanical properties of Si₃N₄ preprocessed ball bearing materials. The materials are classified in three categories, Class 1, 2 and 3, by the specification of their characteristics and microstructures. Figure 13 shows the Weibull plot of the bending strength and other mechanical properties of the fabricated specimens. The shape factor in the Weibull plot of the bending strength was 13, and the average bending strength

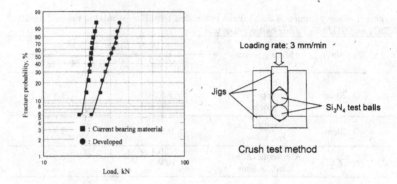

Fig. 14 Crush strength of bearing balls fabricated from the developed Si_3N_4 ceramics.

was 1109 MPa. The fracture toughness and Vickers hardness were 6.7 $MPam^{1/2}$ and 15.4 GPa, respectively. These are higher than the Class 1 values. We prepared balls using the developed material to measure the crushing strength. The crushing strength of this material was about 31.7 N, which is 1.5 times that of the conventional Si_3N_4 ceramics used for bearings (Figure 14). A rolling fatigue test was carried out using 3 mm diameter balls under a pressure of 5.9 GPa. The rolling fatigue lifetime of the developed TiN-nanoparticle-dispersed Si_3N_4 ceramics was longer than 10^7 cycles. Consequently, it was confirmed that the developed material has sufficient mechanical properties for use in the next generation of ceramic bearings.

CONCLUSION

By innovation of the optimum compositions of the α-Si_3N_4-Y_2O_3-Al_2O_3, full densification, high strength and high toughness of Si_3N_4 ceramics have been developed and successfully put into the practical uses such as glow plugs, hot-chambers, turbocharger rotors, wear resistant parts and refractory components for molten Al. Particularly Si_3N_4 bearings have been realized in the first application in 1983. After then it has been recognized that Si_3N_4 bearings are highly advantageous key components in various industries. Furthermore, Si_3N_4 ceramics fabricated by the addition of TiO_2 to the Si_3N_4-Y_2O_3-Al_2O_3 have been widely used for bearing applications

Recently we have confirmed that TiO_2 addition caused grain boundary strengthening of Si_3N_4 ceramics by changing into TiN. Yokohama National University have started the new joint work with Toshiba Materials Co. and JTKT Corp. under the

NEDO contract: "Development of high performance TiN nano-particle dispersed silicon nitride ceramics for rolling bearings". In this project, we have adapted the newly developed nano-TiO_2 dispersion process by the mechano-chemical process. Consequently, we have achieved the targeted values; performance of Class 1 in ISO26602, low aggressiveness to a mating metals, longer lifetime over 10^7cycle under the load of 5.9 GPa by thrust type rolling fatigue test. The evaluations of the developed materials have to be done by bearing assembly.

Acknowledgement
Author acknowledges NEDO and JSPS, Japan, for financial supports and continuous encouragement, and Dr. J. Tatami, Dr. T. Wakihara, Dr. T. Meguro and students, Yokohama National University, Japan, for their much efforts on this items. Furthermore author also thanks for Toshiba Materials Co., Ltd., Japan, and JTEKT Corp., Japan, for their kind support to promote the NEDO Project.

References
[1] J. F. Collins and R. W. Gerby, New Refractory Uses for Silicon Nitride Q13 Reported, *J. Metals*, 7, 612.5 (1955).

[2] G. C. Deeley, J. M. Herbert, and N. C. Moore, Dense Silicon Nitride, *Powder Met.*, 8, 145 (1961).

[3] A. Tsuge, K. Nishida, and M. Komatsu, Effect of Crystallizing the Grain-Boundary Glass Phase on the High Temperature Strength of Hot-Pressed Si_3N_4, *J. Am. Ceram. Soc.*, No. 58, pp. 323-326 (1975).

[4] K. Komeya and F. Noda, Aluminum Nitride and Silicon Nitride for High Temperature Gas Turbine Engines, *SAE* Paper," No. 40237 (1974).

[5] A. Tsuge, H. Inoue and K. Komeya, Grain Boundary Crystallization of silicon Nitride with Material Loss During Heat Treatment, *J. Am. Ceram. Soc.*, 72(10), 2014-16 (1988).

[6] Y. Oyama, and S. Kamigaito, Solid Solubility of Some Oxides in Si_3N_4, Jpn. *J. Appl. Phys.*, No. 10, pp. 1637(1971).

[7] K. H. Jack, and W. I. Wilson, Ceramics Based on the Si-Al-O-N and Related Systems, *Nature Physics Science*. No. 238, 28-29 (1972).

[8] T. Yamada and Y. Kotoku, Commercial production of high purity silicon nitride powder by the imide thermal decomposition method, *Jpn. Chem. Ind. Assoc. Mon.*, 42, 8 (1989).

[9] M. Mitomo, Pressure Sintering of Si_3N_4, *J. Mater. Sci.*, No. 11, 1103-1107 (1976).

[10] K. H. Jack, Review : Sialons and related nitrogen ceramics, *J. Mater. Sci.*, 11, 1135-58 (1976).

[11] S. Hampshire, H. K. Park, D. P. Thompson & K. H. Jack, α-Sialon ceramics, Nature

274, 880 (1978).

[12]A. Rosenflanz and I. W. Chen, Phase Relationships and Stability of α'-SiAlON, *J. Am. Ceram. Soc.*, 84(4), 1025-36 (1999).

[13]H. Kawamura and S. Yamamoto, Improvement of Diesel Engine Startability by Ceramic Glow Plug Start System, *SAE Paper*, No. 830580 (1983).

[14]S. Kamiya, M. Murachi, H.Kawamoto, S. Kato, S. Kawakami and Y. Suzuki, "Silicon Nitride Swirl Chambers for High Power Charged Diesel Engines, " *SAE Paper*, No. 850523 (1985).

[15]H. Hattori, Y. Tajima, K. Yabuta, Y. Matsuo, M.Kawamura and T. Watanabe, "Gas Pressure Sintered Silicon Nitride Ceramics for Turbocharger Application, " *Proc. 2nd International Symposium bon Ceramic Materials and Components for Engines*, pp. 165 (1986).

[16]K. Komeya and H. Kotani, Development of Ceramic Antifriction Bearing, *JSAE Rev.*, No. 7, 72-79 (1986).

[17]H. Takebayashi, K. Tanimoto and T. Hattori, Performance of Hybrid Ceramic Bearing at High Speed Condition (Part 1), *J. Gas Turbine Soc. Japan*, No. 26, pp. 55-60 (1998).

[18]H. Takebayashi, K. Tanimoto and T. Hattori, Performance of Hybrid Ceramic Bearing at High Speed Condition (Prat 2), *J. Gas Turbine Soc. Japan*, No. 26, pp. 61-66 (1998).

[19]W. E. Mandler, Jr., T. M. Yonushonis and K. Shinozaki, Ceramic Successes in Diesel Engines, *presented at 6th International Symposium on Ceramic Materials & Components for Engines*, October 19-23, 1997, Arita, Japan.

[20]K. Komeya, Materials Development and Wear applications of Si_3N_4 Ceramics, *Ceramic Transaction*, Vol. 133, 3-16 (2002).

[21]M. Komatsu, Development of Ceramic Bearing, Ceramics, 39, 633-38 (2004). (in Japanese)

[22]J. Tatami, M. Toyama, K. Noguchi, K. Komeya, T. Meguro and M. Komatsu, Effect of TiO_2 and AlN Additions on the Sintering Behavior of the Si_3N_4-Y_2O_3-Al_2O_3 System, *Ceramic Transactions*, No. 247, pp. 83-86 (2003).

[23]T. Yano, J. Tatami, K. Komeya, and T. Meguro, Microstructural Observation of Silicon Nitride Ceramics Sintered with Addition of Titania, *J. Ceram. Soc. Japan*, No. 109, pp. 396-400 (2001).

[24]J. Tatami, I. W. Chen, Y. Yamamoto, M. Komastu, K. Komeya, D. K. Kim, T. Wakihara and T. Meguro, Fracture Resistance and Contact Damage of TiN Particle Reinforced Si_3N_4 Ceramics, *J. Ceram. Soc. Japan*, No. 114, pp. 1049-1053 (2006).

[25]K. Adachi, K. Kato and N. Chen, Wear Map of Ceramics, *Wear*, No. 203, pp. 291-301 (1997).

[26]S. Zheng, L. Gao, H. Watanabe, J. Tatami, T. Wakihara, K. Komeya and T. Meguro, Improving the Microstructure of Si_3N_4-TiN Composites Using Various PEIs to Disperse Raw TiO_2 Powder, *Ceramics International*, No. 33, pp. 355-359 (2007).

[27] J. Tatami, E. Kodama, H. Watanabe, H. Nakano, T. Wakihara, K. Komeya, T. Meguro and A. Azushima, Fabrication and wear properties of TiN nanoparticle-dispersed Si_3N_4 ceramics, J. Ceram. Soc. Japan, 116, 749-754 (2008).

[28] J. Tatami, H. Nakano, T. Wakihara and K. Komeya, Development of Advanced Ceramics by Powder Composite Process, KONA Powder and Particle Journal, 28, 227-240 (2010).

COMPARISON OF MICROWAVE AND CONVENTIONALLY SINTERED YTTRIA DOPED ZIRCONIA CERAMICS AND HYDROXYAPATITE-ZIRCONIA NANOCOMPOSITES

Mark R. Towler
Inamori School of Engineering, Alfred University, Alfred, NY 14802

Stuart Hampshire, Colin J. Reidy, Declan J. Curran, Thomas J. Fleming
Materials and Surface Science Institute, University of Limerick, Limerick, Ireland

ABSTRACT

Comparisons of microwave and conventional sintering of zirconia and hydroxyapatite (HA) – zirconia bodies were investigated in order to understand how microwave energy may affect the physical and mechanical properties of the materials for use in biomedical applications. Powder compacts of commercial nano-sized ZrO_2, with 2 to 5 mol% Y_2O_3, and mixtures of laboratory synthesised nano-sized HA with 0–10 wt% zirconia were microwave and conventionally sintered at temperatures up to 1450°C for the zirconia and 1200°C for the composites with the same heating profile and a 1h hold time.

Microwave sintering improves physical and mechanical properties of Y_2O_3–doped ZrO_2 ceramics compared with conventional sintering. Compositions containing 2 mol% Y_2O_3 exhibit the greatest improvement due to retention of tetragonal ZrO_2, with higher relative density, 22% increase in Young's modulus and a 165% increase in biaxial flexural strength compared with conventional sintering. Significant grain growth occurred in microwave sintered samples which is thought to be related to enhanced diffusional effects during microwave sintering.

HA-ZrO_2 composites exhibited densities of only 80% with corresponding open porosities. Nanosized ZrO_2 prevents the densification of the HA matrix by effectively pinning grain boundaries and this effect is more pronounced in microwave sintered ceramics. Nanocomposites microwave sintered at 1200°C showed similar strengths to those produced by conventional sintering but with higher volume fraction open porosity. Increased open porosity is considered to be useful for biomedical applications to promote osteo-integration.

INTRODUCTION

Microwave sintering has been developed as a processing method for a range of structural and functional ceramics[1-8] including alumina[1,2], zinc oxide[3] and various electro-ceramics[4], zirconia[5] and non-oxide ceramics and composites[6]. Microwave sintering is known to produce higher densities than would be achieved at the same temperature in a conventional furnace. In addition, from an economic perspective, microwave sintering allows improved properties and better microstructural control with shorter processing times and less energy consumption. The use of microwave processing for dental and bioceramics is under development[9-10] where there may be clear advantages in terms of better densification but with reduced sintering schedules.

Y_2O_3-doped ZrO_2 based ceramics are becoming favoured materials for dental implant applications and these tailored microstructures enhance properties through the use of the tetragonal to monoclinic ZrO_2 phase transformation to allow transformation toughening[11-14]. Typical microstructures for this family of ceramics consist of tetragonal phase precipitates in a cubic ZrO_2 ceramic matrix (PSZ) or simply single phase tetragonal ZrO_2 polycrystals (TZP). Hydroxyapatite (HA) has been shown to be both bio-active[15] and osteo-conductive[16] allowing it to promote new bone growth *in-vivo* without eliciting an immune response. However, the low mechanical strength of HA limits its use to non-load bearing skeletal applications. The use of ZrO_2 as a dispersed phase may improve the mechanical properties of HA[17,18] allowing it to be used in load bearing skeletal applications. However sintering temperatures, even as low as 1100°C, result in the decomposition of HA to tetracalcium phosphate (TTCP) and subsequently tricalcium phosphate (TCP)[19]. These decomposition phases can create a less favourable environment for new bone growth in the body[20] due to increased solubility and increased cytotoxicity. The use of both ZrO_2 particles and microwave sintering may improve the final density, microstructure and strength of HA at much lower sintering temperatures compared with conventional sintering, while reducing decomposition of HA.

Because of the fundamental differences between conventional sintering (radiant heating externally) and microwave sintering (heat generation internally), the aim of the current study was to investigate the effects of both sintering regimes on physical and mechanical properties of (1) Y_2O_3-doped (2-5 mol%) ZrO_2 ceramics and (2) Hydroxyapatite-ZrO_2 composites.

EXPERIMENTAL PROCEDURE

Co-precipitated 3 and 8 mol% Y_2O_3-doped ZrO_2 powders (TZ-3Y and TZ-8Y, Tosoh Corporation, Japan) were mixed with Y_2O_3-free monoclinic powder (TZ-0, Tosoh Corporation, Japan) in order to obtain mixtures with 2, 3, 4 and 5 mol% Y_2O_3. The mixed powders were attrition milled, shell frozen and retrieved using a freeze dryer to obtain a homogeneous mixture which was passed through a 90µm sieve in order to break up soft agglomerates.

Pure hydroxyapatite (HA) powder was produced at a synthesis temperature of 25°C by a precipitation method based on that of Jarcho et al.[21] Full details of the preparation are given in a previous report[10]. HA-ZrO_2 composites were prepared by ball-milling mixtures of precipitated HA and 2 mol% Y_2O_3-doped ZrO_2 (5 and 10 wt%) using alumina milling media for 6 hours.

Particle size was analysed by laser diffraction (Malvern Mastersizer 2000, Malvern, UK) and transmission electron microscopy (TEM, JEM-2011, JEOL Ltd., Japan). Powder compacts were uniaxially pressed at 35MPa and then isostatically pressed at 150MPa to produce disc-shaped green bodies (3mm height, 19mm Ø).

Microwave sintering was performed in a hybrid microwave sintering furnace with a power output of 1.125kW, utilising a SiC susceptor to facilitate hybrid heating. This method minimizes thermal gradients within the sample, whilst providing the ability to heat the samples to a critical temperature beyond which they heat solely due to microwave radiation absorption[22]. Samples were sintered at various temperatures up to 1300°C, held for one hour at the process temperature before being allowed to furnace cool. The temperature profile was recorded and subsequently used to program a conventional resistive element sintering furnace in which comparative firings were performed using the same heating profile.

In order to examine phase assemblages, X-ray diffraction was carried out (Philips X'Pert MPD Pro diffractometer, Philips, The Netherlands) with scans in the range 5°<2θ<80° and a step size of 0.2°, time per step of 20s. The volume fraction of m-ZrO_2 was calculated from the relative intensities of the principle m-ZrO_2 and t/c-ZrO_2 peaks[23] in the range 28-32° 2θ.

Densities of the sintered bodies were determined by an Archimedes method using a Sartorius YDK-01 balance accessory kit. The theoretical density (TD) of zirconia was calculated taking into account the Y_2O_3 content[24] and therefore allowing for the variation in the absolute value of TD with composition.

The elastic modulus was determined using an ultrasonic pulse echo test method based on ASTM E494-10.[25] The longitudinal (V_l) and transverse (V_t) ultrasonic wave velocities were determined on each sample with an accuracy better than 0.1% using two 10 MHz piezoelectric transducers (Ultran Laboratories Inc., Boalsburg, PA, USA). The Young's (E) and shear (G) moduli and Poisson's ratio ( ) were calculated from the following equations:

$$E = (3 V_l^2 - 4 V_t^2) / [(V_l^2 / V_t^2) - 1] \qquad (1)$$

$$G = V_t^2 \qquad (2)$$

$$= (E / 2G) - 1 \qquad (3)$$

Biaxial Flexural Strength (BFS) was determined using 4-point bending tests adapted from ISO 6872[26] using a 5KN load cell (LR50K Load Frame, Lloyd Instruments, West Sussex, UK) with a crosshead speed 0.1mm/min.

RESULTS AND DISCUSSION

COMPARISON OF MICROWAVE AND CONVENTIONAL SINTERING OF ZIRCONIA
Fig. 1 shows the effect of sintering regime (Microwave – MS or Conventional – CS) at 1300°C and Y_2O_3 content on relative density of zirconia.

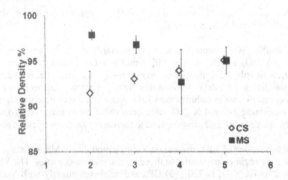

Fig. 1 Effect of Y_2O_3 content and sintering regime (Microwave – MS or Conventional – CS) at 1300°C on relative density of zirconia.

The relative densities of the MS ZrO_2 containing 2 and 3 mol% Y_2O_3 are ~95.5-97%, which is greater than their CS counterparts, the relative densities of which increase linearly with Y_2O_3 content from 90% at 2 mol% Y_2O_3 to 95% at 5 mol% Y_2O_3 content. For 4 and 5 mol% Y_2O_3, densities are similar for both sintering regimes. At lower sintering temperatures (1100 and 1200°C), much higher densities are achieved using microwave sintering but as the sintering temperature is increased the differences in densities between samples from both sintering techniques is reduced considerably. Overall, microwave sintering gave higher densities than conventional sintering, particularly for ZrO_2 with 2 and 3 mol % Y_2O_3.

Fig. 2 shows X-ray diffraction patterns of both MS and CS ZrO_2 ceramics doped with 2-5 mol% Y_2O_3 sintered at 1300°C.

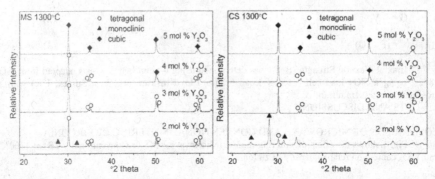

Fig. 2 XRD traces of 2-5 mol% Y_2O_3 doped ZrO_2 samples microwave sintered (MS) and conventionally sintered (CS) at 1300°C.

The CS 2 mol% Y_2O_3 composition consists primarily of monoclinic- (m-) ZrO_2, with only 13.3% tetragonal- (t-) ZrO_2 stabilized. The MS 2 mol% Y_2O_3 sample, however, is almost entirely stabilized to t-ZrO_2, with only 5.5% monoclinic phase present. It is clear that with 2 mol% Y_2O_3, MS is more effective at stabilizing the t-ZrO_2 phase at this "lower" sintering temperature of 1300°C and also that 2 mol% Y_2O_3 is insufficient to stabilize the t-ZrO_2 phase at a conventional sintering temperature of 1300°C. Samples containing 3-5 mol % Y_2O_3 have been stabilised as either tetragonal/cubic ZrO_2 with no trace of m-ZrO_2. Increases in Y_2O_3 content lead to increasing amounts of cubic phase at the expense of t-ZrO_2.

Fig. 3 shows the effect of Y_2O_3 content on Young's modulus of MS and CS ZrO_2 materials. The trends for both sintering regimes are similar to those for the relative densities. The Young's modulus of the MS ZrO_2 with 2 mol% Y_2O_3 is 220 ($\pm$5) GPa and decreases slightly with Y_2O_3 content, all MS ZrO_2 materials possessing elastic moduli in the range 200-220 ($\pm$5-10) GPa. The Young's moduli of the CS ZrO_2 materials increase linearly with increasing Y_2O_3 content, from 180 ($\pm$15) GPa with 2 mol% Y_2O_3 to 205 ($\pm$5) GPa with 5 mol% Y_2O_3. Similar trends are observed for Shear and Bulk Moduli. It is clear that the MS ZrO_2 materials, particularly those with lower Y_2O_3 content, are stiffer than their CS counterparts by 22% at 2 mol% Y_2O_3, as a result of their higher densities. As with density, property values obtained on MS samples are much greater than their CS counterparts at lower sintering temperatures of 1100 and 1200°C.

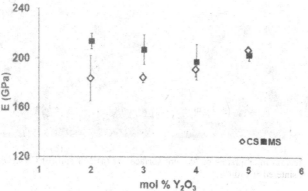

Fig. 3 Effect of Y_2O_3 content and sintering regime (MS or CS) at 1300°C on Young's modulus of ZrO_2 materials.

Fig. 4 shows the effects of Y_2O_3 content on biaxial flexural strength (BFS) of the MS and CS ZrO_2 ceramics. The BFS for ZrO_2 with 2 mol% Y_2O_3 is 800 ($\pm$80) MPa, 165% higher than its CS ZrO_2 counterpart with BFS of 300 ($\pm$200) MPa. For MS samples, BFS decreases to 550 ($\pm$50) MPa with 3 mol% Y_2O_3 and to 420 ($\pm$20) MPa with 5 mol% Y_2O_3. The BFS of the CS ZrO_2 with 2 mol% Y_2O_3 is the lowest of all tested, whereas the strengths of the CS ZrO_2 with 3 to 5 mol% Y_2O_3 are higher in the range ~430-450 ($\pm$30) MPa. The large differences in strength between the MS and CS ZrO_2 with 2 mol% Y_2O_3 is partially due to the differences in density and porosity but is also influenced by the type of ZrO_2 present, the MS sample containing mainly t-ZrO_2 whereas the CS sample contains mainly m-ZrO_2. MS ZrO_2 ceramic doped with 2 mol% Y_2O_3 is the best overall material in terms of density, elastic modulus and, in particular, strength at 800 ($\pm$80) MPa. Scanning electron microscopy shows that in addition to lower porosity, this material has larger grain size than its CS counterpart as shown in Fig. 5, suggesting that there is an enhanced diffusional effect attributable to microwave sintering[3,7].

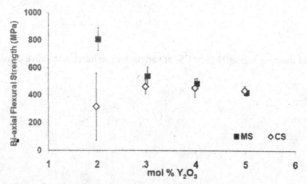

Fig. 4 Effect of Y_2O_3 content and sintering regime (MS or CS) at 1300°C on Bi-axial Flexural Strength (BFS) of ZrO_2 materials.

(a)   (b)

Fig. 5 Scanning electron micrographs of fracture surfaces of (a) MS and (b) CS ZrO_2 containing 2 mol% Y_2O_3 sintered at $1300°C$.

COMPARISON OF MICROWAVE AND CONVENTIONAL SINTERING OF HYDROXYAPATITE-ZIRCONIA COMPOSITES

Fig. 6 shows the effect of sintering regime (MS or CS) at various temperatures and ZrO_2 content on relative density of pure hydroxyapatite (HA)-zirconia composites.

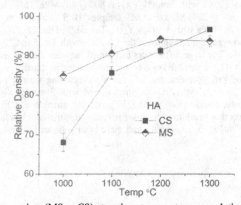

Fig. 6 Effect of sintering regime (MS or CS) at various temperatures on relative density of HA

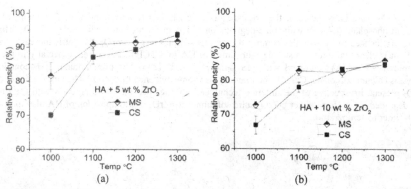

Fig. 7 Effect of sintering regime (MS or CS) at various temperatures on relative density of hydroxyapatite-zirconia composites with (a) 5 wt.% and (b) 10 wt.% ZrO_2.

With no added zirconia, higher densities are achieved using microwave sintering up to 1200°C than for conventional sintering although differences decrease significantly as temperature increases. Fig. 7 shows the effect of sintering regime (MS or CS) at various temperatures and ZrO_2 content on relative density of HA-zirconia composites. With 5 and 10 wt.% ZrO_2, density decreases with zirconia content but is higher for microwave sintering at 1000-1100°C.

Quantitative X-ray analysis was performed in order to assess the amount of HA, ZrO_2 and possible intermediates and degradation products in the sintered bodies and to assess if the different sintering regimes resulted in different phase assemblages. The results for samples sintered at 1200°C are presented in Table I.

Table I. Quantitative phase analysis of Hydroxyapatite-ZrO_2 composites microwave (MS) and conventionally (CS) sintered at 1200°C.

wt % ZrO_2				wt % Phases Present				
MS 1200°C	HA	α-TCP	β-TCP	TTCP	m-ZrO_2	t-ZrO_2	c-ZrO_2	$CaZrO_3$
0	98.7	-	-	1.3	-	-	-	-
5	93.7	0.6	-	2.3	-	0.2	2.8	0.5
10	61.6	-	27.0	3.2	-	0.1	1.7	6.9
CS 1200°C								
0	99.4	-	-	0.4	-	-	-	-
5	91.0	0.6	2.2	-	0.1	0.1	3.5	2.5
10	49.2	0.7	38.8	0.5	-	-	0.6	10.1

There is no evidence of decomposition phases present in the pure hydroxyapatite (HA) except for a very small amount of tetra-calcium phosphate (TTCP) present. In compositions containing 5 wt.% ZrO_2 there is evidence of the presence of cubic zirconia (c-ZrO_2) and a small amount of tricalcium phosphate (β-TCP) with some degradation of HA to form calcium zirconate, $CaZrO_3$. This is more pronounced in CS samples compared with MS samples. As the ZrO_2 content is increased to 10 wt.%, it is clear that much greater decomposition of HA to β-TCP has occurred, especially in the CS samples (49.2% HA) compared with MS samples (61.6% HA). The predominant ZrO_2-containing phase is calcium zirconate, $CaZrO_3$, but there is also some evidence of c-ZrO_2. The small amount of t-ZrO_2 present, irrespective of the sintering technique, indicates that there is a reaction with CaO from the degraded HA to form either c-ZrO_2 solid solution or $CaZrO_3$. Decomposition of HA also increases with sintering temperature.

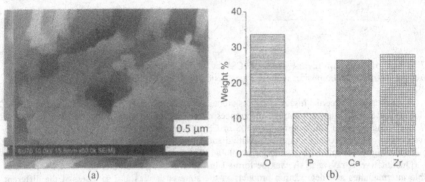

Fig. 8: (a) SEM image of fracture surface of HA matrix showing pore surrounded by ZrO_2 particles; (b) EDX analysis of area around pore in (a).

From scanning electron microscopy, as shown in Fig. 8, there is evidence that ZrO_2 particles segregate at grain interfaces and effectively pin the grain boundaries, inhibit densification and result in increased open interconnected porosity with rougher surfaces.

For mechanical properties, composites microwave sintered up to 1200°C have higher mean biaxial flexure strengths than CS samples but, for both sintering regimes, strengths decrease as ZrO_2 increases, following similar trends as for density. A realistic analysis of the data, which exhibits large error bars, suggests that ZrO_2 content, sintering temperature and regime do not really have a significant effect on strength.

CONCLUSIONS

Microwave sintering (MS) achieved greater densification of Y_2O_3-doped ZrO_2 samples than conventional sintering (CS) at 1300°C. At lower sintering temperatures (1100 and 1200°C), much higher densities are achieved using microwave sintering but as the sintering temperature is increased the differences in densities between samples from both sintering techniques is reduced considerably.

Elastic moduli of MS and CS ZrO_2 ceramics are dependant mainly on the porosity present in the sample, irrespective of the sintering technique or the mol % Y_2O_3 present. Thus E is higher for the higher density MS ZrO_2 samples.

Maximum biaxial flexural strength (BFS) of 800 (±80) MPa was observed for the MS ZrO_2 samples containing 2 mol% Y_2O_3 which was mainly tetragonal ZrO_2 whereas the CS ZrO_2 samples containing 2 mol% Y_2O_3 were mainly monoclinic ZrO_2 and strength was only 300 (±200) MPa. For other levels of Y_2O_3, BFS depends on porosity, phase assemblage and microstructure (grain size).

Introduction of zirconia particles into hydroxyapatite leads to lower densification with increasing ZrO_2 content as a result of pinning of grain boundaries and also degradation of HA and formation of tri-calcium phosphate (TCP) and calcium zirconate, $CaZrO_3$, regardless of the sintering method used. This results in higher porosities and lower mechanical properties.

ACKNOWLEDGEMENTS

This work was funded under the Enterprise Ireland Commercialisation Fund Technology Development Programme (EI/TD/2006/0328).

REFERENCES

[1] M. A. Janney, H. D. Kimrey, M. A. Schmidtand, and J. O. Kiggans, "Grain Growth in Microwave Annealed Alumina," *J. Am. Ceram. Soc.*, **74** [7], 1675–81 (1991).

[2] K. H. Brosnan, G. L. Messing, and D. K. Agrawal, "Microwave Sintering of Alumina at 2.45 GHz," *J. Am. Ceram. Soc.*, **86** [8], 1307–12 (2003).

[3] J. Binner, J. Wang, B. Vaidhyanathan, N. Joomun, J. Kilner, G. Dimitrakis and T. E. Cross, "Evidence for the Microwave Effect During Annealing of Zinc Oxide," *J. Am. Ceram. Soc.*, **90** [9], 2693-97 (2007).

[4] C. Y. Fang, C. A. Randall, M. T. Lanagan, and D. K. Agrawal, "Microwave Processing of Electroceramic Materials and Devices," *J. Electroceram.*, **22**, 125–30 (2009).

[5] S. Charmond, CP. Carry, D. Bouvard. "Densification and microstructure evolution of Y-Tetragonal Zirconia Polycrystal powder during direct and hybrid microwave sintering in a single-mode cavity", *J. Euro. Ceram. Soc.*, **30** [6], 1211-21 (2010).

[6] S. Zhu, W. Fahrenholtz, G. Hilmas, S. Zhang, E. Yadlowsky & M. Keitz, "Microwave sintering of a ZrB_2-B_4C particulate ceramic composite", *Composites Part A: Appl. Sci. Manufact.*, **39** [2], 449-53 (2008).

[7] J. Wang, J. Binner, B. Vaidhyanathan, N. Joomun, J. Kilner, G. Dimitrakis and T. E. Cross, "Evidence for the Microwave Effect During Hybrid Sintering," *J. Am. Ceram. Soc.*, **89** [6], 1977–84 (2006).

[8] D. Agrawal, J. Cheng, H. Peng, L. Hurt and K. Cherian, "Microwave energy applied to processing of high-temperature materials," *Am. Ceram. Soc. Bull.*, **87** [3], 39-44 (2008).

[9] X. Wang, H. Fan, Y. Xiao, X. Zhang, "Fabrication and characterization of porous hydroxyapatite/β-tricalcium phosphate ceramics by microwave sintering," *Mater. Letts.*, **60**, 455-8 (2006).

[10] D. J. Curran, T. J. Fleming, M. R. Towler and S. Hampshire, "Mechanical Properties of Hydroxyapatite – Zirconia Composites Produced by Two Different Sintering Methods," *J. Mater. Sci.-Mater. in Medicine*, **21** [4], 1109-20 (2010).

[11] R. C. Garvie, R. H. J. Hannink and R. T. Pascoe, "Ceramic Steel?" *Nature (London)*, **258**, 703–4 (1975).

[12] P. F. Becher and M. V. Swain, "Grain-Size-Dependent Transformation Behavior in Polycrystalline Tetragonal Zirconia," *J. Am. Ceram. Soc.*, **75** [3], 493–502 (1992).

[13] R. H. J. Hannink, P. M. Kelly and B. C. Muddle, "Transformation toughening in Zirconia-Containing Ceramics," *J. Am. Ceram. Soc.*, **83** [3], 461-87 (2000).

[14] B. Basu, J. Vleugels and O. Van der Biest, "Toughness tailoring of yttria-doped zirconia ceramics," *Mater. Sci. Eng. A*, **380** [1-2], 215-21 (2004).

[15] M. Wang. 'Developing bioactive composites for tissue replacement', *Biomaterials*, **24**[13] 2133-51 (2003).

[16] K. Anselme, B. Noel, B. Flautre, MC. Blary, C. Delecourt, M. Descamps, P. Hardouin. "Association of porous hydroxyapatite and bone marrow cells for bone regeneration," *Bone*, **25** [2], 51-4 (1999).

[17]M. R. Towler, I . R. Gibson, S. M. Best, "Novel processing of hydroxyapatite-zirconia composites using nano-sized particles," *J. Mater. Sci. Letts.*, **19**, 2209-11 (2000).

[18]Z. Evis and R. H. Doremus, "Hot-pressed hydroxylapatite/monoclinic zirconia composites with improved mechanical properties," *J. Mater. Sci.*, **42**, 2426–31 (2007).

[19]D. Bernache-Assollant, A. Ababou, E. Champion and M. Heughebaert, "Sintering of calcium phosphate hydroxyapatite $Ca_{10}(PO_4)_6(OH)_2$ I. Calcination and particle growth," *J. Euro. Ceram. Soc.*, **23** [2], 229-41 (2003).

[20]L. L. Hench, "Bioceramics: From concept to clinic," *J. Am. Ceram. Soc.*, **74** [7], 1487-1510 (1991).

[21]M. Jarcho, C. H. Bolen, M. B. Thomas, J. Bobick, J. F. Kay and R. H. Doremus, "Hydroxylapatite synthesis and characterisation in dense polycrystalline form," *J. Mater. Sci.*, **11** [11], 2027-35 (1976).

[22]P. D. Ramesh, D. Brandon and L. Schachter, "Use of partially oxidized SiC particle bed for microwave sintering of low loss ceramics", *Mater. Sci. & Eng. A – Struct. Mater.* **266** [1-2], 211-20 (1999).

[23]H. Toraya, M. Yoshimura and S. Somiya, "Calibration Curve for Quantitative Analysis of the Monoclinic-Tetragonal ZrO_2 System by X-Ray Diffraction," *J. Am. Ceram. Soc.*, **67** [6], 119-21 (1984).

[24]R. P. Ingel and D. Lewis III, "Lattice Parameters and Density for Y_2O_3-Stabilized ZrO_2", *J. Am. Ceram. Soc.*, **69** [4], 325-32 (1986).

[25]A. Díaz, S. Hampshire, J-F. Yang, T. Ohji and S. Kanzaki, "Comparison of Mechanical Properties of Silicon Nitrides with Controlled Porosities Produced by Different Fabrication Routes," *J. Am. Ceram. Soc.*, **88** [3], 698–706 (2005).

[26]ISO. Dentistry - Ceramic Materials, ISO6872 (2008).

RECENT DEVELOPMENTS IN HIGH THERMAL CONDUCTIVITY SILICON NITRIDE CERAMICS

You Zhou [a], Kiyoshi Hirao [a], Hideki Hyuga [a], Dai Kusano [a,b,c]
[a] *National Institute of Advanced Industrial Science and Technology (AIST), Nagoya 463-8560, Japan*
[b] *Department of Frontier Materials, Nagoya Institute of Technology, Nagoya 466-8555, Japan*
[c] *Japan Fine Ceramics Co. Ltd., Sendai 981-3203, Japan*

ABSTRACT

In this report, we briefly review the background and some important technical issues in developing high thermal conductivity silicon nitride ceramics. Then, we report our recent studies on preparing silicon nitride ceramics via a route of sintering of reaction-bonded silicon nitride (SRBSN), where green compacts composed of a high purity silicon powder and a small amount of sintering additives were firstly nitrided in a nitrogen atmosphere and then sintered at higher temperatures to attain full densification. By minimizing the oxygen content in the starting powder and controlling the nitridation and sintering conditions, silicon nitride ceramics possessing high thermal conductivity (>100 W/(mK)) and high bending strength (>800 MPa) were fabricated. The concurrent attainment of both high thermal conductivity and good mechanical properties makes these silicon nitride ceramics promising candidates for high performance electrical substrate materials.

INTRODUCTION

Energy and environment-related problems are serious social issues. In order to save energy as well as to reduce the emission of carbon dioxide, energy sources tend to shift from fossil fuel to electric power, hence highly efficient use of electric power becomes extremely important. Power devices are key technologies for this purpose. Power devices conduct direct conversion and control of electric power by means of semiconductors, which can drastically save energy compared to traditional systems. In recent years, power supply and packing density of power modules are rapidly increasing as their application field expands, particularly in the automobile industry.

In order to guarantee the stable operation of the power module, heat release technology in the module becomes very important. So far, AlN has been used as a substrate for power devices. In general, a thick metal plate as an electrode is directly bonded to a ceramic substrate via high temperature brazing as illustrated in Fig. 1. With increase of the power density in the module, micro-cracks in the ceramic caused by the large difference in thermal expansion coefficient of metal and ceramic become serious problem. Such situation leads to strong demands for insulating substrates possessing better mechanical properties, and thereby attention is turned to silicon nitride.

Figure 2 summarizes the properties of commercial ceramic substrates. The thermal conductivities of commercially available silicon nitrides are less than 90 W/(mK), though their strengths are more than twice as high as those of aluminum nitrides. An important issue in silicon nitride is, therefore, to increase thermal conductivity without degradation of mechanical properties. So far some researchers have estimated the intrinsic thermal conductivity of -Si_3N_4 crystal based on solid-state physics.[1,2] It is anticipated that the intrinsic thermal conductivity of silicon nitride is higher than 200 W/(mK).[3]

Being different from the ideal single crystal, a sintered silicon nitride is composed of -Si_3N_4 grains and two-grain boundaries as well as secondary phases with low thermal conductivity. In addition, inside the grains, there exist a variety of imperfections such as point, line, planer defects.[4,5] These defects scatter phonons to decrease thermal conductivity of -Si_3N_4 crystal. Similar to the case of AlN ceramics, oxygen impurity dissolved in Si_3N_4 lattice can drastically decrease the thermal conductivity of silicon nitrides.[4,6]

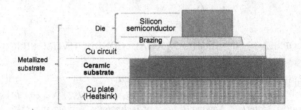

Figure 1. Structure of a power module using a ceramic substrate

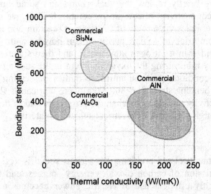

Figure 2. Properties of commercially available ceramic substrates.
(Data were collected through trade catalogs or from websites of
Japanese ceramics manufacturing companies.)

In order to reduce these negative parameters, high thermal conductivity Si_3N_4 is generally fabricated with the following points in mind: (1) High purity and fine Si_3N_4 starting powder is employed in order to reduce impurities. (2) Concurrent addition of rare earth oxides and alkaline earth oxides is conducted. The former plays a role of decreasing lattice oxygen, and the latter is needed for assisting densification. With respect to alkaline earth element, use of $MgSiN_2$ as one of the sintering additives was found to be effective for improving thermal conductivity.[7] (3) Enhancement of grain growth is one important factor for improving thermal conductivity because it can decrease the number of grain boundaries as well as the lattice oxygen via solution re-precipitation process.[3]

In practice, gas pressure sintering (GPS) with long sintering time over several tens of hours is necessary in order to achieve high thermal conductivity over 100 W/(mK). Figure 3 shows a relation between thermal conductivity and bending strength of GPSed silicon nitrides. Considerable grain growth is needed in order to improve thermal conductivity over 100W/(mK), which results in drastic decrease in strength. For the commercial silicon nitride powders, even the highest grade powder contains more than 1 wt% of impurity oxygen, one third of which is lattice oxygen, which makes it difficult to fabricate high thermal conductivity Si_3N_4 with high strength via the GPS method.

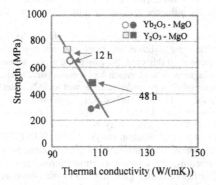

Figure 3. Relation between thermal conductivity and bending strength of gas-pressure sintered silicon nitrides with Yb_2O_3-MgO or Y_2O_3-MgO as sintering additives. Sintering was done at 1900 °C for various times.

In order to overcome this problem, our research group has carried out studies on preparing high-thermal-conductivity silicon nitrides via a route of sintering of reaction-bonded silicon nitride (SRBSN).[8-10] The process has advantages such as: (1) The starting Si powder is cheaper than the high purity Si_3N_4 powder which is used in the conventional sintering process; (2) high dimensional accuracy due to low shrinkage after post-sintering; (3) It is possible to carry out the whole process from nitridation to post-sintering process without exposing the compacts to the air, which is favorable for controlling the oxygen content in Si_3N_4. In the following sections, we report some recent experimental results obtained in our study.

EXPERIMENTAL

A high purity Si powder (Kojundo Chemical Laboratory Co. Ltd, Saitama, Japan) with a mean particle size (d_{50}) of 8.5 m, a total metallic impurities content of less than 0.01 wt% and oxygen content of 0.28 wt% was used as the starting powder for the SRBSN process. High purity Y_2O_3 (>99.9% purity, Shin-Etsu Chemical Co. Ltd., Tokyo, Japan) and MgO (>99.9%, UBE Industries Co. Ltd., Yamaguchi, Japan) were used as sintering additives. The amount of sintering additives in the Si compact was determined so as to let the fully-nitrided RBSN have a nominal composition of Si_3N_4 : Y_2O_3 : MgO = 93 : 2 : 5 in molar ratio.

The Si and sintering additives were mixed in methanol using a planetary mill in a Si_3N_4 jar with Si_3N_4 balls. The rotational speed was 250 rpm and milling time was 2 h. After vacuum drying and sieving, about 18 g of mixed powder was uniaxially pressed in a 45 mm x 45 mm stainless-steel die and then cold-isostatically pressed at a pressure of 300 MPa. The Si compact, embedded in a BN powder bed, was placed inside a BN crucible which was then put into a larger graphite crucible. Both nitridation and post-sintering were conducted in a graphite resistance furnace. The green compact was nitrided at 1400 °C for 8 h under a nitrogen pressure of 0.1 MPa. Then, by further heating up the furnace and introducing more nitrogen, post-sintering of the nitrided RBSN compact was carried out at 1900 °C under a nitrogen pressure of 1 MPa. Holding times at the sintering temperature ranged from 1 to 48 h in order to investigate the influence of sintering time on the properties of the SRBSN materials.

For the sake of comparison, materials were also prepared by the conventional gas pressure sintering (GPS) method using same sintering additives and same sintering parameters as those used for post-sintering of the SRBSN materials. In this case, the raw powder was a fine α-Si_3N_4 powder (E-10, UBE Industries Co. Ltd., Yamaguchi, Japan), which had a mean particle size of 0.2 μm and contained 1.2 wt% of oxygen. The sintered silicon nitrides were termed as SSN materials thereafter.

Bulk density was measured by the Archimedes method using deionized water as an immersion medium. Theoretical density (TD) was calculated according to the rule of mixtures. Relative density was given by the ratio of bulk density and theoretical density. Phase identification for the nitrided and post-sintered samples was performed by X-ray diffraction analysis (XRD) (RINT2500, Rigaku, Tokyo, Japan) with Cuk radiation of 40kV/100mA. Microstructures of fracture surfaces of the nitrided and the post-sintered samples were observed using a scanning electron microscope (SEM) (JSM-6340F, JEOL, Tokyo, Japan) equipped with a field emission gun. Test beams with dimensions of 4 mm x 3 mm x 36 mm were sectioned from the sintered specimens and ground with a 400-grit diamond grinding wheel. The beams were tested in a four-point bending jig with an outer span of 30 mm, an inner span of 10 mm, and at a crosshead speed of 0.5 mm/min.

Thermal diffusivity was measured by the laser flash method (Model TC-7000, ULVAC, Yokohama, Japan). Disk specimens with a dimension of 10 mm in diameter and 3 mm in thickness were cut from the sintered materials. The disks were sputter-coated with a 60 nm thick film of gold to prevent direct transmission of the laser, followed by a subsequent coating of black carbon to increase the amount of energy absorbed. Thermal conductivity (κ) was calculated according to the equation

$$\kappa = \rho C_p a \qquad (1)$$

where ρ C_p and α are bulk density, specific heat and thermal diffusivity, respectively.

RESULTS AND DISCUSSION

The measured thermal conductivity values of the sintered reaction bonded Si_3N_4 (SRBSN) and the gas pressure sintered Si_3N_4 (SSN) materials which were sintered at 1900 °C for various times are shown in Fig. 4. When sintered for the same times, the SRBSN materials always had higher thermal conductivities than the SSN counterpart materials. And, when the sintering times were longer, the difference in the thermal conductivities between the SRBSN and SSN materials became larger. After being sintered for 48 h, the SRBSN material attained a thermal conductivity over 140 W/(mK).

In Fig. 5, the thermal conductivities of the SRBSN and the SSN materials are plotted against their bending strengths. It can be seen that the SRBSN materials had a much better balance between thermal conductivity and bending strength than the SSN materials. The better properties of thermal conductivity and strength of the SRBSN materials could be attributed to the lower oxygen content of the Si powder which was used as the starting powder for the preparation of SRBSN materials than that of the Si_3N_4 starting powder for the SSN materials. The lower oxygen content would make the SRBSN materials attain a higher thermal conductivity than the SSN materials even when they were sintered under same conditions (temperature, time and sintering additive).

Figure 6 shows the SEM micrographs of the fracture surfaces of the SRBSN materials sintered at

1900 °C for 6 and 24 hours. They showed some common features: elongated and faceted grain shapes, a bimodal microstructure in which a small fraction of large grains dispersed in a majority of small grains. With increasing sintering time, the microstructures became coarser. The fracture mode of the materials was predominantly intergranular fracture, suggesting that they might have high fracture toughness. Indeed, our preliminary experimental results did reveal that the fracture toughness of these materials were rather high. The detailed results will be reported in the near future.

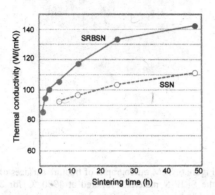

Figure 4. Thermal conductivity of the sintered reaction bonded Si₃N₄ (SRBSN) and the gas-pressure sintered Si₃N₄ (SSN) sintered at 1900 °C for various times under 1 MPa N₂.

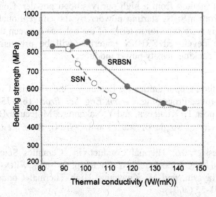

Figure 5. Relation between thermal conductivity and bending strength for the sintered reaction bonded Si₃N₄ (SRBSN) and the gas-pressure sintered Si₃N₄ (SSN) sintered at 1900 °C for various times under 1 MPa N₂.

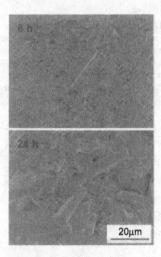

Figure 6. SEM micrographs of fracture surfaces of
the SRBSN materials sintered at 1900 °C for 6
and 24 hours.

SUMMARY

Silicon nitride ceramics with both high thermal conductivity and good mechanical properties
were prepared by a SRBSN method from a high purity silicon powder. Compared with the silicon
nitride ceramics prepared from a Si_3N_4 starting powder by the conventional gas-pressure sintering
method, the SRBSN materials were superior in terms of the balance between thermal conductivity and
mechanical properties. The developed SRBSN materials will be promising for applications as
advanced substrates for high packing density power modules.

REFERENCES

[1]J.S. Haggerty and A. Lightfoot, Opportunities for Enhancing the Thermal Conductivities of SiC and
Si_3N_4 Ceramics through Improved Processing, Ceram. Eng. Sci. Proc., 16, 475-87 (1995).
[2]N. Hirosaki, S. Ogata, C. Kocer, H. Kitagawa, and Y. Nakamura, Molecular Dynamics Calculation of
the Ideal Thermal Conductivity of Single-Crystal - and -Si_3N_4, Phys. Rev. B, Article No. 134110
(2002).
[3]K. Hirao and Y. Zhou, Session 16: Thermal Conductivity, Ceramics Science and Technology –
Properties (Volume 2), edited by R. Riedel and I-W. Chen, pp. 667-696, Wiley-VCH (2010).
[4]K. Hirao, K. Watari, H. Hayashi, and M. Kitayama, High Thermal Conductivity Silicon Nitride
Ceramics, MRS Bull., 26, 451-55 (2001).
[5]M. Kitayama, K. Hirao, A. Tsuge, K. Watari, M. Toriyama, and S. Kanzaki, Thermal Conductivity of
-Si_3N_4: I. Effects of Various Microstructural Factors, J. Am. Ceram. Soc., 82, 3105-12 (1999).

[6]M. Kitayama, K. Hirao, K. Watari, M. Toriyama, and S. Kanzaki, Thermal Conductivity of -Si$_3$N$_4$: II, Effect of Lattive Oxygen, *J. Am. Ceram. Soc.,* **83**, 1985-92 (2000).

[7]H. Hayashi, K. Hirao, M. Toriyama, S. Kanzaki, and K. Itatani, MgSiN$_2$ Addition as a Means of Increasing the Thermal Conductivity of -Silicon Nitride, *J. Am. Ceram. Soc.,* **84**, 3060-62 (2001).

[8]X.W. Zhu, Y. Zhou, K. Hirao, and Z. Lences, Processing and Thermal Conductivity of Sintered Reaction-Bonded Silicon Nitride. I: Effect of Si Powder Characteristics, *J. Am. Ceram. Soc.,* **89**, 3331-39 (2006).

[9]X.W. Zhu, Y. Zhou, K. Hirao, and Z. Lences, Processing and Thermal Conductivity of Sintered Reaction-Bonded Silicon Nitride. (II) Effect of Magnesium Compound and Yttria Additives, *J. Am. Ceram. Soc.,* **90**, 1684-92 (2007).

[10]Y. Zhou, X.W. Zhu, K. Hirao, and Z. Lences, Sintered Reaction-Bonded Silicon Nitride with High Thermal Conductivity and High Strength, *Int. J. Appl. Ceram. Technol.,* **5**, 119-26 (2008).

MICROSTRUCTURE MAPS FOR UNIDIRECTIONAL FREEZING OF PARTICLE SUSPENSIONS

Rainer Oberacker[a], Thomas Waschkies[a,b], Michael J. Hoffmann[a]

[a] Institute for Applied Materials -Ceramics in Mechanical Engineering-, Karlsruhe Institute of Technology (KIT), 76131 Karlsruhe, Germany
[b] now with Fraunhofer Institute for Non-Destructive Testing (IZFP), 66123 Saarbrücken, Germany

ABSTRACT

The structure formation in water based ceramic suspensions during unidirectional freezing was experimentally investigated over a wide range of solidification velocities. Supplementary experiments with polystyrene suspensions were carried out to extend the range of particle sizes. Depending on particle size, solids loading and solidification velocity, planar, lamellar or isotropic growth of the ice crystals leads to different types of microstructure. The results are summarized in a microstructure map as a first estimate for the role of the process parameters in microstructure formation.

1 INTRODUCTION

Ice-templated structure formation is a manufacturing method for macroporous ceramics which received increasing interest during the last years. The process starts with (unidirectional) freezing of the liquid suspension medium of ceramic slurries. During solidification, the suspension medium and the ceramic particles separate, forming clusters with a highly dense particle packing and clusters without particles or with a much diluted particle concentration. During subsequent sublimation, the latter leaves macropores, while the former generates regions of well packed particle arrangements. The well packed regions can be sintered to a desired level of microporosity or even to full density, while the macropores remain. A lot of efforts were put to this production technique which shows the ability in using a wide variety of materials and dispersion liquids [1,2,3,4,5,6]. A recent review[7] summarizes these methods.

Water is commonly used as dispersion liquid and will exclusively be discussed further in this paper. The total porosity can be easily controlled by the solids loading of the slurry, as the fraction of suspension medium determines the macropore volume in the green state[9,10]. It is also known, that increasing solidification velocities lead to decreasing spacing of the ice crystals, while increasing solids loadings lead to increasing ice crystals spacing[8]. In sintered samples, a dense layer is sometimes found in the bottom region of unidirectional solidified samples where high cooling rates prevail and therefore high solidification velocities. It is assumed, that this dense layer results from frozen structures with homogenously packed particles and finely dispersed isotropic ice crystals. At somewhat lower solidification velocities, a lamellar microstructure is formed consisting of parallel ice crystals separated by particle lamellae. This is the desired morphology for manufacturing materials with directional pore channels. Lamellae formation is a consequence of the interaction between the solidification front and the particles. The spacing of the lamellae increases as the solidification rate decreases. However, the obtainable maximum spacing has not yet been discussed in literature. Nevertheless, one can expected, that a lower boundary exists for the solidification velocity at which lamellae formation takes place. If the solidification front moves very slowly, the particles are mobile enough to move away from the solidification front. In earlier publications Rempel and others[11,12,13,14,15] studied the interaction of particles with a moving ice-liquid interface and found a transition from pushing to trapping of the particles by the planar solidification front with increasing solidification velocities.

The aim of the present paper is to link these different findings and to derive the processing conditions under which different microstructure morphologies can be realized. For this purpose the microstructure transition regions for water based suspensions were experimentally analyzed for different particle sizes, solids loadings and solidification velocities. In order to avoid artefacts, the microstructures were investigated in frozen or in green state. With these methods, the microstructure development was investigated from very slow (< 1 μm/s) to very high (> 100 μm/s) solidification velocities.

2 EXPERIMENTAL

Different water based suspensions were used for the experiments. Table I gives the type and mean particle size of the raw materials. The alumina slurries were mixed with a dispersant (Dolapix CE 64, Zschimmer&Schwarz GmbH, Germany) in distilled water and ball-milled for about 4h. Solids loading of the suspensions ranged from 5 to 30 vol.%. Experiments with larger particle sizes were conducted with water based polystyrene suspensions with a solids loading of 5 vol.%. The low density of polystyrene helps to avoid sedimentation artefacts. Freezing was done in cylindrical acrylic glass moulds with 30 mm diameter for the alumina suspensions and 8 mm diameter for the polystyrene.

Table I. Particle systems used for the present investigations

Powder			d_{50} (μm)
Alumina Suspensions	AA-03	Sumitomo Chemical Co., LTD., Japan	0.25
	AA-3		3.3
	CT3000SG	Almatis GmbH, Germany	0.8
Polystyrene Suspensions		Test particles, BS-Particle GmbH, Germany	2.2
			4.1
			6.6
			15

The exploration of planar ice growth requires very low solidification velocities. For this purpose, acrylic glass moulds were glued on a thin copper plate and placed on a pre-cooled plate of a freeze-dryer (Type Sublimator 400K, Zirbus technology, Germany). In a first step, a bottom layer of several millimetres of pure water was frozen inside the mould to reduce the heat flow and thus the solidification velocity. In a second step the plate of the freeze-drier was adjusted between 0.5 to 1 °C and the suspensions were poured onto the planar ice. The microstructure development over time was investigated by a Long-Distance-Microscope (K2, ISCO-OPTIC GmbH, Germany).

Single-side as well as double-side cooling experiments were applied in the region of lamellar ice growth. For single-side cooling experiments, freezing was done on a copper cooling plate of a freeze-dryer, pre-cooled to constant temperatures. The top of the acrylic glass moulds were open, so that the upper surface of the slurry was exposed to the atmosphere at room temperature. In single side cooling, the solidification velocity decreases with increasing progress of the solidification front which corresponds to the sample height. The double-side cooling experiments were carried out in a custom-made double side cooling setup (Figure 1) Cooling occurs from two copper rods at the bottom and the top of the mould. The temperatures of the rods can be changed during freezing, in order to establish constant freezing velocities in the range from 5 μm/s to 30 μm/s over a sample height of 4.5 cm[16]. The solidification velocity for both setups was derived from image sequences of the optically visible solidification front, taken with a macro lens.

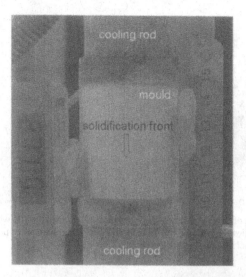

Figure 1. Double side cooling setup (DS-Setup). With the aid of pre-calculated temperature profiles, constant freezing velocities in a range from 5 to 30 μm/s over a total sample height of 45 mm can be realized[16]

The microstructure was analyzed either from freeze dried and resin infiltrated samples, or from direct observation of the frozen structure inside a cryostage of a SEM (Leica Stereoscan 440, Leica Cambrige LTD., England). In both cases, the samples were sectioned at several heights perpendicular to the freezing direction.

To realize very fast solidification velocities, small slurry droplets were put onto a copper plate pre-cooled to – 196 °C by liquid nitrogen. The average freezing velocity was estimated from the total solidification time and the droplet dimensions. After freezing, the microstructure was analyzed from fracture surfaces of the droplet in the Cryo-SEM.

3 RESULTS
3.1 Low solidification velocities: Transition between planar and lamellar ice growth

The freezing process for water based suspensions is commonly discussed only for solidification velocities above 1 μm/s[7]. Investigations for solidification velocities below this level are lacking. Therefore, solidification experiments were carried out with 0.8 μm sized alumina powder and 6.6 μm polystyrene particles. Solids loading c_V of the alumina slurries was varied from 5 up to 30 vol.%. The solidification velocity was varied in order to determine the transition velocity v_T where the planar ice growth changes to a lamellar ice growth.

Figure 2A shows the solidification progress over time for a slurry with $c_V = 5$ vol.% and a solidification velocity of 0.06 μm/s. The solidification front remains planar, pushing the ceramic particles ahead of the front. Only a few particles are engulfed by the ice, visible from the small dark speckles in the image. With increasing solidification velocity, the planar solidification front becomes unstable (Figure 2B). For a solidification velocity of 1.2 μm/s, ice lamellae are formed which push the ceramic particles into the interlamellar spaces, visible as dark lines growing from bottom to top.

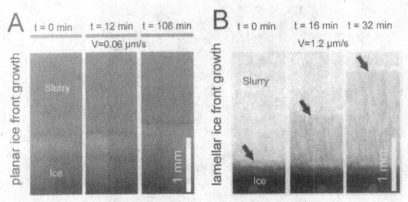

Figure 2. Transition of the ice morphology from planar (A) to lamellar (B) with increasing solidification velocities for a slurry with 5 vol.% alumina. Images are taken perpendicular to the ice front and solidification starts from bottom.

The transition velocity v_T is shown in Figure 3. For 5 vol.% alumina solids loading, v_T is about 0.35 µm/s. It decreases to about 0.05 µm/s for 30 vol.% solids loading. The dependence in this concentration interval is almost linear. The transition velocity for polystyrene suspensions with 5 vol.% solids loading and 6.6 µm particle size is about 0.2 µm/s, which is significantly below the transition velocity for alumina suspensions with the same solids content.

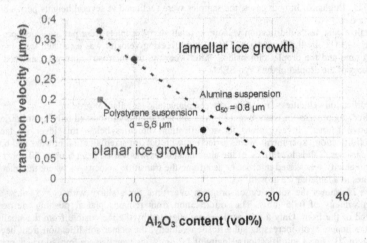

Figure 3. Relationship between the transition velocity v_T from a planar to a lamellar ice morphology and solids content for alumina and polystyrene suspensions.

3.2 Intermediate solidification rates: Lamellar ice growth of alumina suspensions

For solidification velocities beyond 0.5 μm/s lamellar ice growth was observed for all of the investigated alumina suspensions. The relationship between lamellae spacing, solidification velocity and solids loading was studied from single-side, as well as from double-side cooling experiments. The solidification velocity was varied between about 2 μm/s and 40 μm/s at solids loadings between 5 to 30 vol.%. Fig. 4 shows an example of the cross sections near the bottom (A) respectively top (B) of samples after unidirectional freezing by single side cooling. The difference in solids loading results in a clear difference in pore volume. The solidification velocities and lamellae spacing are plotted in the graphs included in the figure.

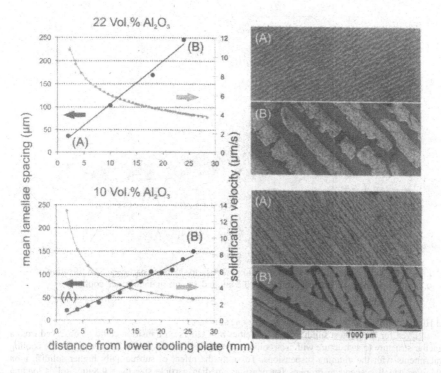

Figure 4. lamellae spacing over sample height during single side cooling experiments with solids loadings of 22 and 10 vol.%. Mean particle size 0.8 μm, cooling plate temperature -10 °C.

Most important results are summarized in Fig. 5; detailed data and a quantitative analysis of the complete data set are subject of a forthcoming paper. The measurements show, that the empirical dependence of the lamellae spacing λ on the solidification velocity v can be described by a power law ($\lambda \propto v^{-1}$), as already reported by Deville et al.[8]. The influence of particle size on the lamellae development is not clearly identified in literature. Our experiments with the different mean particle sizes of 0.3, 0.8 and 3.3 µm indicate a shift to a wider lamellae spacing with increasing particle size. An increase in particle size of a factor of 10 leads approximately to a doubling of lamellae spacing.

An increase in solids loading results also in a wider lamellae spacing. This can be seen from the experiments with the 0.8 µm particles, represented by the grey shaded area in Fig. 5. The lower boundary represents a suspension with 5 vol.% solids loading, the upper boundary is valid for 30 vol.% solids loading. The dashed line represents 10 vol.% solids loading.

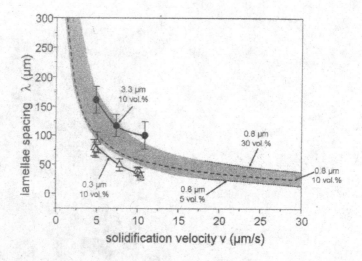

Figure 5. Lamellae spacing vs. solidification velocity for alumina particles. Solidification of 0.8 µm suspensions in single side cooling, 0.3 µm and 3.3 µm in double side cooling

3.3 High solidification rates: Transition to isotropic structures

Even for the highest solidification velocities of 40 µm/s no transition could be observed from a lamellar structure to a structure with isotropic ice crystals, neither for double nor for single side cooling experiments with the alumina suspensions. To study the effect of substantially higher solidification velocities, small suspension droplets (suspensions: median particle size $d_{50} = 0.8$ µm, solids loading $c_v = 10$ and 20 vol.%) were placed on a pre-cooled copper plate cooled to - 196°C. Freezing occurs very rapid under such conditions. Depending on the droplet size, cooling rates of 250 to 700 µm/s could be realized. Figure 6A shows the side view of a fully frozen droplet in the Cryo-SEM. At higher magnification (Figure 6B) the formation of ice lamellae growing from the interface at the cooling plate can be seen. The average lamellae spacing was about 6 to 10 µm for a solidification velocity of 250 µm/s and about 3 µm for solidification velocities of 700 µm/s. No significant difference in lamellae spacing could be found for the two investigated solid loadings.

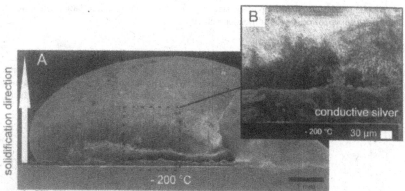

Figure 6. SEM image of a unidirectional frozen droplet of alumina suspension with 20 vol.% solids loading. The solidification velocity is ~ 250 μm/s

These findings show that the small alumina particles are mobile enough to assort to particle lamellae between the growing ice crystals even at maximum solidification velocities. To study the behaviour of coarser particles, experiments with water based polystyrene suspensions were performed. Four different suspensions with particle sizes of 2.2, 4.1, 6.6 and 15 μm were unidirectional frozen by the DS-device (Fig. 1) at freezing velocities between 3 and 20 μm/s. Except of the 2.2 μm particle suspensions, all systems undergo a change from lamellar to isotropic, when a critical solidification velocity is exceeded (Figure 7B). For a particle size of 4.1 μm, the microstructure is lamellar for solidification velocities of 8 and 10 μm/s, respectively, but isotropic for 20 μm/s. For a particle size of 6.6 μm, the microstructure transition occurs by changing the solidification velocity from 5 to 8 μm/s. The 15 μm particle suspensions form lamellae at 1 μm/s, but are entrapped at 8 μm/s. The 2.2 μm particle suspension solidifies to a lamellar microstructure up to 10 μm/s. A droplet experiment at ~ 200 μm/s, however, resulted in an isotropic microstructure. Figure 7A gives an overview of the complete experimental results. The transition velocity from a lamellar microstructure to an isotropic microstructure is drastically shifted to smaller solidification velocities with increasing particle sizes.

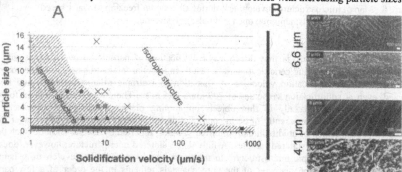

Figure 7. (A) Influence of particle size on the microstructure development for alumina and polystyrene suspensions with particle sizes of 0.8 μm to 15 μm. (B) SEM images of

microstructure development for particle sizes of 4.1 and 6.6 µm for solidification velocities below and above the critical solidification velocity.

4 DISCUSSION AND CONCLUSIONS

The main objective of the experiments was to identify the microstructure morphologies, which can be created by directional solidification of water based particle suspensions and to determine the critical processing conditions for morphology transitions. The controlling processing parameters are the particle size, suspension solids loading and the solidification velocity. The major experimental findings are summarized in Fig. 8 in form of a microstructure map.

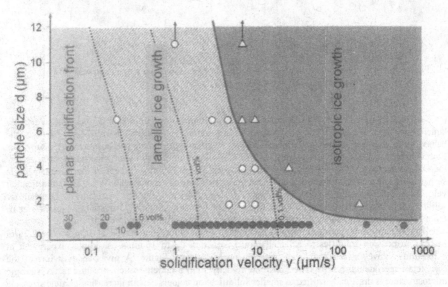

Figure 8. Map of microstructure development during directional freezing of water based suspensions. Filled symbols: alumina; open symbols: polystyrene.

An important finding is that directional solidification of particle suspensions, as they are typically used for engineering ceramic materials, results in lamellar microstructures over the complete range of practical solidification velocities. A transition to isotropic microstructures could not be observed, even at solidification velocities of 700 µm/s (2.5 m/h). The cooling rates needed for such velocities can be realized only in thin objects on cooling plates pre-cooled to extremely low temperatures. This seems apparently to be in contradiction to the observation of a dense layer in the fast cooled bottom region of unidirectional solidified samples, reported by several authors. Such dense layers were observed with sintered samples. A fully dense sintered microstructure, however, does not necessarily need an isotropic microstructure in the green state, but can also develop from lamellar particle packings, when the diameter of the pore channels remains in the order of a few particle diameters.

A lower limit for the lamellar morphology is given by the transition velocity v_T, below which a planar growth of the ice front is observed. The transition velocity derived from the model experiments decreases strongly with increasing solids loading of the suspension, as visible from the data points in Fig. 8. Under practical conditions, a planar growing ice front leads to an increase of solids loading at the ice – suspension interface and thus to a steadily decreasing transition velocity. This takes place at solidification velocities of less than 1 mm/h, which can hardly be controlled under practical conditions. It follows, that solely lamellar microstructures can be expected within the range of technical processing conditions for engineering ceramics,.

The model experiments with polystyrene suspensions were used to derive rough estimates for the border lines of the morphology fields in the case of an increasing particle size. Although it has to be taken into account that the solids loading of the polystyrene suspension was only 5 vol.% and that the interparticle forces may be different from the alumina system, some general trends can be derived. The planar to lamellar transition velocity decreases slightly with increasing particle size. The lamellar to isotropic transition velocity, which was only observed for the polystyrene suspensions, decreases strongly with particle size. Both are caused by decreasing particle mobility. In consequence, the interval of solidification velocities leading to lamellar microstructures is much wider for practical particle sizes on a micrometer- or even sub-micrometer level, but much narrower for coarse particles of several micrometers in diameter.

The planar to lamellar boundaries for very dilute suspensions for 1.0 or 0.1 vol.% solids loading (dashed lines) are a qualitative extrapolation of the experimental results from the alumina suspensions. They give an indication for the behaviour of dilute suspensions. The solidification rates necessary for forming lamellar microstructures have to be increased with decreasing solids loading. The lamellar microstructure field narrows with increasing dilution. This leads to crucial consequences for the quantitative aspects of microstructure. The lamellae spacing which can be controlled via the solidification velocity can be varied over a wide range with particle suspensions typically applied in engineering ceramics. Lamellae spacing between a few micrometers and several hundreds of micrometers can be realized. This is valid over a wide interval of solids loading, the parameter which controls the macropore volume fraction of the final material. Coarse particles and very dilute suspensions limit the feasible range of spacing dimensions.

Unidirectional solidification of particle suspensions is therefore a dedicated method for the production of engineering ceramics with aligned pore channels in the sub-millimetre range. The mechanisms of lamellae formation are still under discussion and models of the freezing process are far away from practical applicability. Therefore, the development of new models is needed. Our microstructure map in Fig. 8 gives a first estimate of the boundaries for the process parameters which lead to the desired lamellar channel formation. The quantitative relations between lamellae spacing and process parameters derived from the experiments will be treated in a forthcoming paper.

ACKNOWLEDGEMENT
The work was funded by Deutsche Forschungsgemeinschaft (DFG) under grant OB 104/11.

REFERENCES
[1] K. Araki, J. Halloran, New freeze-casting technique for ceramics with sublimable vehicles, *J. Am. Ceram. Soc.* **87** (10), 1859–1863 (2004).
[2] K. Araki, J. Halloran, Porous ceramic bodies with interconnected pore channels by a novel freeze casting technique, *J. Am. Ceram. Soc.* **88** (5) 1108–1114, (2005).
[3] K. Lu, C. Kessler, R. Davis, Optimization of a nanoparticle suspension for freeze casting, *J. Am. Ceram. Soc.* **89** (8) 2459–2465 (2006).

[4] S. Deville, E. Saiz, R. Nalla, A. Tomsia, Freezing as a path to build complex composites, *Science* **311** 515–518 (2006).

[5] T. Moritz, H.-J. Richter, Ice-mould freeze casting of porous ceramic components, *J. Eur. Ceram. Soc.* **27** (16) 4595–4601 (2007).

[6] Y. Chino, D. Dunand, Directionally freeze-cast titanium foam with aligned, elongated pores, *Acta Mater.* **56** (1) 105–113, (2008).

[7] S. Deville, Freeze-casting of porous ceramics: A review of current achievements and issues, *Adv. Eng. Mater.* 10 (3) 155–169 (2008).

[8] S. Deville, E. Saiz, A. Tomsia, Ice-templated porous alumina structures, *Acta Mater.* **55** (6) 1965–1974 (2007).

[9] S. Deville, E. Saiz, A. Tomsia, Freeze casting of hydroxyapatite scaffolds for bone tissue engineering, *Biomaterials* **27** (32) 5480–5489 (2006).

[10] T. Waschkies, A. Mattern, R. Oberacker, M. Hoffmann, Macro pore developement during freeze-casting of water based alumina suspensions, *cfi/Ber. DKG* **86** E41–E46 (2009).

[11] A. Rempel, M. Worster, The interaction between a particle and an advancing solidification front, *J. Cryst. Growth* **205** (3) 427–440 (1999).

[12] A. Rempel, M. Worster, Particle trapping at an advancing solidification front with interfacial-curvature effects, *J. Cryst. Growth* **223** (3) 420–432 (2001).

[13] G. Bolling, J. Cissé, A theory for the interaction of particles with a solidifying front, *J. Cryst. Growth* **10** (1) 56–66 (1971).

[14] J. Pötschke, V. Rogge, On the behaviour of foreign particles at an advancing solid-liquid interface, *J. Cryst. Growth* **94** (3) 726–738 (1989).

[15] C. Körber, G. Rau, M. D. Cosman, E. G. Cravalho, Interaction of particles and a moving ice-liquid interface, *J. Cryst. Growth* **72** (3) 649–662 (1985).

[16] T. Waschkies, R. Oberacker, M. Hoffmann, Control of lamellae spacing during freeze casting of ceramics using double-side cooling as a novel processing route, *J. Am. Ceram. Soc.* **92** (s1) S79–S84 (2009).

INTERACTIONS OF Si$_3$N$_4$-BASED CERAMICS IN WATER ENVIRONMENT UNDER SUB-CRITICAL CONDITIONS

Pavol Šajgalík[1], Dagmar Galusková[2], Miroslav Hnatko[1], Dušan Galusek[2]
[1]Institute of Inorganic Chemistry, Slovak Academy of Sciences, Bratislava, Slovakia
[2]Vitrum Laugaricio – Joint Glass Center of the IIC SAS, TnU AD, FChPT STU, and RONA, j.s.c., Trenčín, Slovakia

ABSTRACT

Two kinds of tests, both under static and *quasi*-dynamic conditions, were applied in order to understand dissolution mechanism of Si$_3$N$_4$ and sialon ceramics under subcritical conditions in neutral aqueous solutions. The Si$_3$N$_4$ ceramics corroded preferentially by congruent dissolution of silicon nitride matrix grains irrespective of the used corrosion solution, with initial dissolution rates of 15.0 and 20.1 mmol·m^{-2}·h^{-1} in de-ionized water and sodium chloride solution, respectively, as determined at 290°C. Three fold increase of dissolution rate of sialon in aqueous sodium chloride solution at 290°C was observed in comparison to dissolution in de-ionized water. At 150 °C the influence of corrosion on strength is negligible. At higher temperatures the decrease of strength was dramatic.

INTRODUCTION

Current interest in the materials chemically resistant in H$_2$O environment under sub/supercritical conditions, favor constantly research on chemical stability of ceramic materials. Supercritical water (water at temperatures above the critical temperature and typically at high pressure) is a challenging environment attracting scientists and engineers because of its importance in many natural situations (such as geothermal formation of minerals) and technological applications (as a working heat transfer fluid in conventional and nuclear power plants and as a solvent of highly toxic material like chlorinated dioxins). A key aspect of each technology concerns identification of materials, and component designs suitable for handling the high temperatures and pressures and aggressive environments present in many processes.

Depending upon the particular application, the supercritical water (SCW) process environment may be oxidizing, reducing, acidic, basic, nonionic, or highly ionic[1]. The fact that water at typical SCW operating conditions loses the ability to solvate charged (polar) species, however, leads to opposite deductions that corrosion is typically more severe in subcritical water (< 370 °C and density above 200-300 kg/m^3) than supercritical water. Maximum point of corrosion is in fact often exhibited in the region just below the critical point (<374°C and 22.1MPa), where the temperature is high enough to promote fast kinetics, and concentrations of corrosion-causing species such as halide ions (and their corresponding dissociation constants) are still sufficiently high for reaction to occur[2]. The most common materials studied in respect to corrosion resistance in SCW systems are nickel-based alloys and austenitic stainless steels[2-6].

Only a few ceramic materials are considered to be resistant against oxidizing high-temperature solutions. These are some zirconia- and alumina-based materials, whereas all non-oxide ceramics are expected to be converted into their oxides or dissolved rapidly. However, in the world of structural ceramics, the silicon-based ceramics (SiC, Si$_3$N$_4$) offer a wide range of thermal and mechanical properties that classify them as popular materials for a wide variety of applications. Authors[7-11] who studied interactions of Si-based ceramics concluded that water oxidizes Si$_3$N$_4$ actively under hydrothermal conditions. On the other hand, selection of appropriate operating conditions and presence of passivation surface layer can in many cases enhance the degree of resistance to chemical attack and so the life-performance of material.

This article discusses the behavior of silicon nitride prepared with the addition of yttria as sintering additive, and of a sialon ceramic in contact with an aqueous solution of sodium chloride and

45

de-ionized water under conditions close to, or within the sub-critical region (150-300°C)[12]. Sodium chloride is a common by product of the treatment of chlorinated wastes in SCW oxidation process units and is known as an aggressive corrosive agent of car parts due to winter road maintenance, or in pumps for hot high salinity solutions used in units for desalination of seawater. The static and *quasi*-dynamic test procedures were developed to obtain relevant information for evaluation of the chemical durability of tested ceramics as potential materials for mentioned applications. Corrosion in de-ionized water as the reference corrosion medium was carried out simultaneously in order to reveal potential influence of the ionic strength. The emphasis was put on the analysis of specific elements dissolved in the solution due to interactions between the material and the corrosion medium, rather than on determination of the weight change alone, with complementary detailed study of the corrosion layer composition (phase and chemical). By combination of all mentioned techniques a set of data was obtained, which facilitated determination of the kinetic parameters of corrosion, together with broad analysis of corroded material and formed secondary products.

EXPERIMENTAL PART
Materials
    Two materials, Si$_3$N$_4$ and sialon were selected to study the corrosion behavior of non-oxide, silicon nitride-based materials. The silicon nitride specimens were prepared from the powder SN-E10 (UBE Industries, Ltd., Japan) with 5 wt% of Y$_2$O$_3$ (4N label, Pacific Industrial Development Corporation) as the sintering aid. For preparation of sialon with the nominal composition Si$_2$Al$_4$O$_4$N$_4$ the following commercial powders were used: 47.9 wt% of α-Al$_2$O$_3$ (Martoxide PS-6, Martinswerk, Germany); 32.9 wt% of α-Si$_3$N$_4$ / LC12-S (H.C. Starck, Germany) and 19.2 wt% of AlN / grade C (H.C. Starck, Germany). The powder batch was attrition milled in isopropanol for 4 h. The powders were dried under infrared lamp and sieved respectively through a 25 μm and 71 μm sieve to prepare Si$_3$N$_4$ and sialon powders with required granulometry. Blocks with dimensions of 50 x 50 x 5 mm^3 prepared by hot pressing at 1750 °C and pressure 30 MPa under nitrogen pressure of 0.1 MPa were cut and ground into rectangular bars 3 × 4 ×50 mm^3, their faces polished, and edges chamfered. The bars were then ultrasonically cleaned in acetone for 15 min, rinsed with distilled water, and dried for 3 h at 110 °C.

Corrosion experiments under static conditions
    De-ionized water (conductivity 25±5 10^{-4}S/m) used as the reference medium was prepared in the reverse osmosis purification system (CHEZAR, Bratislava, Slovak Republic). The 0.5 mol·L^{-1} NaCl solution was prepared by dissolving 29.96±0.03 g of NaCl (MERCK, Darmstadt, Germany) in 1000 mL of de-ionized water. The bars were placed in teflon-lined pressure corrosion reactors with the inside volume 26 cm^3 filled with corrosion liquid and heated in laboratory drying oven. Both static and *quasi* – dynamic tests were carried out at the temperature of 290 °C with maximum duration of the test 480 h. Additional static tests at temperatures 150 and 200°C were applied in order to obtain the data for determination of kinetic parameters, and for calculation of the activation energies. The ratio between the sample surface and the volume of corrosive liquid (*S/V*) was held constant at 0.76±0.03 cm^{-1}. Fresh specimens were used for each test. After the test the samples were removed from the reactors, rinsed with distilled water, dried and weighed. In parallel, the tests without the ceramic specimens were carried out for given time interval, providing the eluate labeled as a *blank*. The concentration of the particular element detected in the *blank* was subtracted from the concentration determined in the eluates taken after the corrosion tests with the ceramic material.

Quasi – dynamic test

To avoid oversaturation of solution with regard to precipitation of corrosion products, which could act as a passivation layer, *quasi*-dynamic test conditions were also applied. Due to high temperatures used typical flow-through experiment with fresh corrosion medium continuously flowing around the tested material could not be performed. Therefore an alternative arrangement has been used, where every 22 h the reactors were cooled down, the corrosive solution was removed and fresh solution was added to the reactor. The test was carried out in two reactors in parallel at the temperature of 290 °C, and with the same specimens for the whole duration of test. The *S/V* ratio was identical to the static tests. The content of released elements including the total concentration of ammonia-bound nitrogen (both in ionised and in molecular form) were determined in each eluate.

NL_i value calculations

The amount of an element released into solution was expressed in terms of the value NL_i in [g/m^2] according to the equation (1), which is often applied in studies of chemical durability of glasses[13]:

$$NL_i = \frac{c_i}{w_i \dfrac{S}{V}} \quad (1)$$

where c_i is the concentration of an element i in solution in [mg·L^{-1}], S is the surface of the ceramic material in contact with corrosive solution in [m^2], V is the volume of the corrosive solution in [m^3], and w_i is the mass fraction of the element i in the ceramic material. This way the amount of the element released into the solution is normalized with respect to its content in corroded material. Identical NL_i values of all elements indicate congruent dissolution of the material. The variation of the NL_i values indicates either preferential dissolution and release of some elements, or precipitation of corrosion products and depletion of the solution of some elements.

Data acquired from the *quasi* – dynamic tests at the respective time of corrosion t were recalculated according to the equation (2):

$$NL_i^t = \frac{c_i}{w_i \dfrac{S}{V}} + NL_i^{t-\Delta t} \quad (2)$$

The equation (2) includes cumulative effect of the amount of dissolved elements in the studied time interval Δt.

Methods of analysis

The contents of metallic elements (Si, Al, Y) released into the solution were determined with the use of the Inductively Coupled Plasma Atomic Emission Spectroscopy (ICP AES, VARIAN Vista MPX) with radial viewing, equipped with V-groove nebuliser and Sturman-Masters spray chamber. A simple internal standardization technique with beryllium as an internal standard was utilized for multi-element analysis of the eluate solutions containing 0.5 mol·L^{-1} NaCl, to eliminate possible errors introduced by the solution with high salt content[14].

For quantification of the total ammonia nitrogen (both in ionised and molecular form) in eluates the spectrophotometric method with sodium nitroprusside (UV-VIS-NIR spectrometer Varian Carry) was employed according to the standard STN ISO 7150-1.

Corrosion surfaces and polished cross sections of corroded specimens were examined by scanning electron microscopy (SEM/EDX, Carl Zeiss SMT, model EVO 40 HV) at accelerating beam voltage 20 kV. Corrosion products formed at exposed surfaces were investigated with the use of X-ray diffractometer (XRD, Bruker D8 DISCOVER) specially designed for measurement of thin layers, in the 2θ interval $20 - 80°$ using CuKα radiation.

The five specimens were used for determination of bending strength after corrosion. The tests were carried out in four point flexure arrangement with the inner and outer span of 40 mm and 20 mm and the cross head speed 0.5 mm·min⁻¹ with the use of the equipment Lloyd LR5K+. The fracture surfaces were examined immediately after flexure tests with the use of the SEM Jeol 7000F.

RESULTS AND DISCUSSION
Static experiments

The time dependences of the weight change of both studied ceramic materials, normalized with respect to the corroded surface area for the tests carried out both in de-ionized water and 0.5 mol·L⁻¹ NaCl solution at 290°C are summarized in Figure 1 (a) and (b). Unlike for sialon specimens, in de-ionized water rapid loss of weight of Si₃N₄ specimens was observed at the temperature of 290°C up to 48 h of the test later followed by a steady state until the end of the experiment. Considering sialon specimens, nearly no weight change was observed for de-ionized water, while for NaCl solution approximately linear decreasing trend was preserved in the whole studied time interval (Figure 1 (b)). In the first 96 hours of dissolution similar weight loss-time dependence of the Si₃N₄ material for both corrosion solutions (Figure 1 (a)) was documented then followed by rapid change related to corrosion in 0.5 mol·L⁻¹ NaCl solution at 290°C.

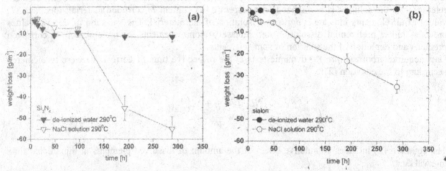

Figure 1. Weight loss measured for Si₃N₄ (a) and sialon (b) specimens in de-ionized water and aqueous sodium chloride solution at 290°C.

The preliminary considerations based on weight loss measurements could classify sialon as promising and chemically resistant material, which could be applied under studied conditions and in the studied temperature interval. However, the chemical analysis of eluates did not support the expectations. Various factors, especially inhomogeneous dissolution in some places (pitting corrosion) with simultaneous precipitation of reaction products and formation of passivation layer, or random cracking off the passivation layer might influence the weight loss - time dependences significantly. We therefore consider the amounts of dissolved elements in corrosion solution for evaluation of corrosion rates and mechanisms, as will be discussed later.

Reaction of Si_3N_4 with water proceeds according to the equation 3 yielding ammonia and silicon dioxide as the reaction products[15].

$$Si_3N_4 + 3(n+2)H_2O + 4H^+ \Rightarrow 3SiO_2 \cdot nH_2O + 4NH_4^+ \qquad (3)$$

Ammonia determined in eluates after corrosion of both studied materials at 290°C was recalculated to the normalized amount of nitrogen and compared to the amount of dissolved silicon (Figure 2 (a)). No specific influence of the selected corrosion media on dissolution of silicon nitride ceramics could be observed (Figure 2 (a)). Identical NL values of N and Si determined in the corrosion mediums indicate preferential degradation of Si_3N_4 matrix grains. Yttrium was not considered because concentrations measured both in de-ionized water and sodium chloride solution were meeting limits of detection (0.01 $mg \cdot L^{-1}$) for applied analytical method. However, higher $NL(N)$ values were detected for dissolution of sialon in NaCl solution at 290°C, which was nearly 8 times higher than for dissolution in de-ionized water (Figure 2 (b)) at the same temperature.

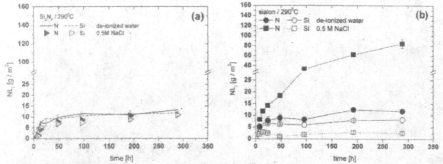

Figure 2. Time dependence of $NL(N)$ values obtained for Si_3N_4 (a) and sialon (b) specimens dissolved under static conditions at 290°C in both corrosion media.

In the both corrosion solutions after the test with sialon specimens, already in the first hours of dissolution lower NL values of dissolved silicon were measured in comparison to NL values of ammonia nitrogen (Figure 2 (b)) and with very low NL of aluminium, ranging from only 0.05 to 0.1 g/m^2 (not shown in Figure 2 (b)). This indicates that the saturation with respect to Si and Al was attained already in the early stage of the dissolution reaction as will be discussed in more detail below.

The SEM analysis of the Si_3N_4 specimen corroded for 288 h at 290°C (Figure 3 (a, b)) indicates the presence of newly formed SiO_2 phase, cristoballite. The layer of precipitates was found to form already after 48 h of the test in de-ionized water, remaining compact during whole time interval studied (Figure 3(a)). The retardation of the corrosion process had been also observed based on weight loss data (Figure 1(a)). The surface of Si_3N_4 ceramics corroded in NaCl solution was covered with a layer of corrosion products, morphologically, and from the point of view of their phase composition, identical to that found at the surface of the Si_3N_4 material corroded in de-ionized water. The X-ray diffraction of corrosion products removed from the surface of the Si_3N_4 specimen revealed the presence of crystalline cristoballite (PDF 76-941). The main difference observed when compared to corrosion in de-ionized water was extensive cracking and peeling off of the layer of precipitates (Figure 3 (b)).

The crystal-like phases covering the surface of sialon ceramics corroded 288 h at 290°C (Figure 3 (c, d)) are of different morphology depending on the corrosion medium applied. The XRD analysis of corrosion products removed from the specimen's surface assigned to the tests in de-ionized water, revealed the presence of α aluminium oxide (PDF 89-7717) together with the $Si_2Al_4O_4N_4$ (PDF 48-1617) sialon: both phases were identical to that detected by XRD in the underlying uncorroded matrix. The presence of layer of corrosion products covering completely the surface of sialon specimens explains the steady state achieved already in an early stage of the corrosion process in de-ionized water (Figure 2 (b)), as well as negligible weight change as seen in Figure 1 (b). The retardation of the corrosion process in the latter hours of the test observed in de-ionized water was not recorded in case of corrosion tests in NaCl solution (Figure 2 (b)). De-lamination of the precipitated layer (Figure 3(d)) might prevent or delay the attainment of the steady state and may be also responsible for the abrupt weight loss observed after 96 h (Figure 1(b)). The XRD analysis of the corrosion products revealed the presence of $NaAl_3Si_3O_{10}(OH)_2$ (paragonite, PDF-24-1047) together with α aluminium oxide (corundum, PDF-71-1123) and boehmite (PDF-1-1283).

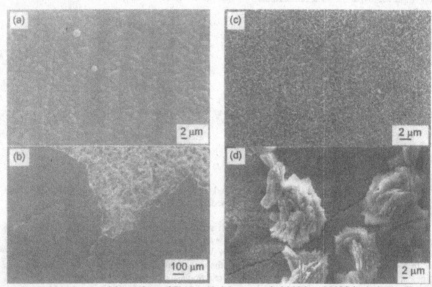

Figure 3. SEM image of the surface of Si_3N_4 and sialon corroded 288 h at 290°C in de-ionized water (a, c) and in 0.5 mol·L⁻¹ (b, d), respectively.

It was documented the corrosive behavior of anions in aqueous environment under subcritical conditions[2,3]. Halides, especially chloride have a remarkable influence on the corrosion of stainless steels and alloys[5,6]. In this respect the chloride anion could be responsible for destruction of the protecting oxide layer covering both studied ceramic materials corroded 288 h in aqueous sodium chloride solution at 290°C.

Quasi-dynamic experiments

In closed system practically every process will meet saturation problems. *Quasi*-dynamic condition where therefore applied in order to eliminate formation of corrosion products at the temperature 290°C. The kinetic parameters were calculated based on an element, in our case nitrogen, which is not incorporated into any secondary phase detected by X-ray diffraction, and remains in solution in the form of dissolved ammonium ion. The cumulative NL values of leached nitrogen and silicon calculated according to (2) and their time dependences in de-ionized water and in 0.5 mol·L^{-1} NaCl solution are respectively shown in Figure 4 (a) and (b). Evaluation of experimental data obtained for the Si_3N_4 ceramics from *quasi*-dynamic tests at 290°C showed only negligible influence of corrosion media on the dissolution process. Speaking in terms of the weight loss, the amount of Si_3N_4 dissolved in de-ionized water and in 0.5 M NaCl solution was 0.005 g·cm^{-2} and 0.006 g·cm^{-2}, respectively, with the relative standard deviation (RSD) 7 %. The Si_3N_4 material dissolves congruently with similar $NL(N, Si)$ values irrespective of the corrosion solution used. Higher magnification images of attacked places, Figure 5, revealed needle-shaped imprints of dissolved Si_3N_4 grains in residual glassy phase. These observations together with the chemical analysis of the eluate suggest that yttrium silicate oxynitride grain boundary glass phase is, at least under the applied test conditions, resistant to attack by the used corrosion solutions.

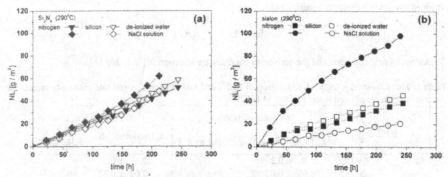

Figure 4. Time dependences of $NL(N)$ and $NL(Si)$ leached from Si_3N_4 (a) and sialon (b) specimen in de-ionized water and in aqueous sodium chloride solution under *quasi*-dynamic conditions at 290°C.

Simultaneous congruent release of Si and N to de-ionized water was observed in eluates from corrosion tests of the sialon ceramics, while Al, detected in solution in significantly lower amounts, re-precipitates in the form of hydroxid-oxides as confirmed by X-ray diffraction where it was identified together with α-Al₂O₃ (corundum) phase, likely originating as a residuum from the original, uncorroded, material.

The influence of Na⁺ and Cl⁻ ions on dissolution of sialon ceramics is evident. The $NL(N)$ values were nearly three times higher comparing to $NL(N)$ calculated for dissolution in de-ionized water, and can be in part related to increased ionic strength of the corrosion solution. In addition, formation of precipitation products, AlOOH and paragonite $NaAl_3Si_3O_{10}(OH)_2$ different from those detected in de-ionized water, deplete the solution of Al and Si promoting further dissolution of sialon (Figure 4(b)).

Figure 5. SEM image of Si₃N₄ specimen analysed after corrosion in *de-ionized water* at 290°C. The Si₃N₄ grains are etched away.

The rate constants of dissolution at the temperature <290°C and at 290°C were determined from the linear parts of the experimental $NL(N)$ (mol·m^{-2}) vs time dependences obtained from the analysis of the data acquired respectively from static and *quasi*-dynamic tests. The activation energies were obtained using the Arrhenius equation (4) (Figure 6),

$$\ln k = \ln A - E_a /(RT) \tag{4}.$$

The values of rate constants and the activation energies are summarized in Table I.

Table I: The activation energy of dissolution of Si₃N₄ and sialon ceramics and rate constants measured for corrosion in de-ionized water and 0.5 M NaCl.

sample		De-ionized water		0.5 mol/L NaCl	
		k_w [mmol·m^{-2}·h^{-1}]	E_w [kJ·mol^{-1}]	k_c [mmol·m^{-2}·h^{-1}]	E_c[kJ·mol^{-1}]
Si₃N₄	290°C	15.0 ± 0.2		20.1 ± 0.7	
	200°C	0.69 ± 0.07	73.6 ± 8	0.63 ± 0.03	69.3 ± 16
	150°C	0.11 ± 0.01		0.16 ± 0.01	
Sialon	290°C	10.6 ± 0.3		27.0 ± 2	
	200°C	1.8 ± 0.3	29.5±10	-	-
	150°C	1.4 ± 0.2		-	

(-) *not determined*

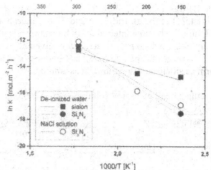

Figure 6. Arrhenius plot of *ln k* vs *1000/T* for sialon and Si_3N_4.

The apparent activation energy of dissolution of Si_3N_4 in de-ionized water and in 0.5 M NaCl solution is 73.6±8 and 69.3±16 kJ/mol, respectively. Similar values of E_a (73-110 kJ/mol) were estimated by Sato[16], and Somiya[10], who concluded that dissolution of Si_3N_4 ceramics under hydrothermal conditions (below 300°C) is controlled by surface chemical reactions. In contrary, Yoshimura concluded that the activation energies at the level 73-84 kJ/mol as determined in his work[8] correspond to diffusion of H_2O in amorphous silica. Based on the literature data it is therefore not possible to make unambiguous conclusions on the controlling mechanism. Very low concentrations of yttrium determined in the corrosion medium suggest a high chemical durability of yttrium oxy-nitride glassy phase in the tested Si_3N_4 ceramics. The corrosion rate 14.2 ± 1 $mmol \cdot m^{-2} \cdot h^{-1}$ determined by Nickel[7] for Si_3N_4 powder corroded at 300°C show excellent agreement, within error levels, with the findings obtained in this study, 15.0 ± 0.2 $mmol \cdot m^{-2} \cdot h^{-1}$.

Apparent activation energy of dissolution of sialon in de-ionized water was 29.5±10 kJ/mol and initial dissolution rate at 290°C was 10.6 $mmol \cdot m^{-2} \cdot h^{-1}$, which is comparable to dissolution rate determined for Si_3N_4 ceramics. Nearly threefold increase of dissolution rate of sialon in aqueous sodium chloride solution at 290°C was observed in comparison to dissolution in de-ionized water, achieving the value 27.0 $mmol \cdot m^{-2} \cdot h^{-1}$.

Geochemical Code Modeling

Modeling of the solution composition using PHREEQC code was performed in order to compare the experimental concentrations of ammonia and silicon present in de-ionized water after the corrosion experiments on Si_3N_4 at 290°C. PHREEQC version 2 is a program[17] written in the C programming language. It is a computer program designed to perform a wide variety of low-temperature aqueous geochemical calculations that are based on an ion-association aqueous model and has capabilities for speciation and saturation-index calculations, as well as many others. Detailed overview and description of the software can be found in the PHREEQC manual[17]. Windows version besides Graphical User Interface includes database files. For our purpose we have used LLNL database with thermodynamic data for a set of minerals.

Based on the results from *quasi*-dynamic experiments the amount of dissolved ceramic material had been calculated with respect to particular time interval. Using PHREEQC code[17] appropriate amount (in moles) of the material with the nominal composition equivalent to the real material, $Si_{2.03}Y_{0.04}N_{2.71}O_{0.07}$, was dissolved in 0.01 kg of de-ionized water with pH 5.63 at the temperature 290°C. Satisfactory agreement between experimentally measured solution composition and calculated solution composition (Table II) was achieved when considering a set of secondary alteration products as are cristobalite (alpha), cristobalite (beta), tridymite and SiO$_2$ (amorphous). These were allowed to precipitate during the calculation. In the time interval of 48 h solution became saturated with respect to cristobalite (alpha) as the only phase, which precipitated under given conditions. The calculated results show that the pH is roughly 2.0 lower than pH measured at laboratory condition. Silicon was present in the solution in dissolved form as silicic acid and nitrogen as ammonia.

Table II. Comparison of calculated and measured solution composition under the static test conditions in de-ionized water at 290°C.

	Measured			Phreeqc code calculation			
Time [h]	pH 25°C	Si [mg·L⁻¹]	N [mg·L⁻¹]	pH 25°C	pH 290°C	Si [mg·L⁻¹]	N [mg·L⁻¹]
24	9.65	255±36	224±30	10.26	7.66	228	152
48	9.65	356±48	306±33	10.35	7.74	343	228
96	9.69	440±20	329±27	10.41	7.79	453	304
192	9.84	456±14	311±47	10.46	7.87	456	380

Modeling of solution composition in case of sialon involves some difficulties. First of all elimination of formation of corrosion products under the *quasi*-dynamic condition was not achieved for sialon specimen. Secondly solution can be oversaturated with respect to a number of minerals bearing aluminium in their structure, where the suitable thermodynamic data are required and so careful examination of appropriate database is necessary.

Nevertheless based on all above discussed results, we assume that under given conditions of corrosion experiments Si-N bonds in Si$_3$N$_4$ and Si-N together with Al-N bonds in sialon are destroyed preferentially. Hydrolysis and adsorption reactions gradually result in breaking silicon-nitrogen bonds to release ammonium, which dissolves in solution changing pH from 6.2±0.3 to 9.7±0.1. In such solution silicon is present in dissolved form as Si(OH)$_4$. In closed system equilibrium is rapidly achieved and solution becomes saturated with respect to the new secondary phases, which precipitate on the surface of corroded material and form passivation layer. A sketch of possible interactions of the ceramic material with aqueous solution is illustrated on the Figure 7.

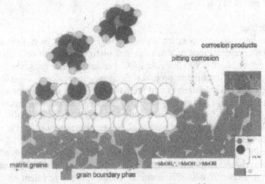

Figure 7. Scheme of the ceramic material interactions with aqueous environment.

Influence on bending strength

Figure 8 (a) summarises the changes in bending strength as the result of corrosion in NaCl solution. For Si_3N_4 the temperature is a crucial factor for both corrosion media. At 150 °C the influence of corrosion on strength is negligible. At higher temperatures the decrease of strength was dramatic, decreasing from 700 to 290 MPa after 288 h at 290 °C. In un-corroded Si_3N_4 specimens the fracture usually originated from processing defects with diameter typically at the level of 40 μm present in the volume of material. Independently of applied corrosion solution, in corroded bars the fracture always originated at the surface, showing corrosion damaged zones hundreds of micrometers in size (pitting corrosion) (Figure 8 (b)).

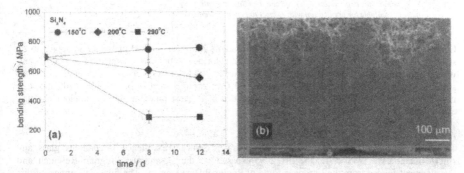

Figure 8. Bending strength time dependence for Si_3N_4 specimens after corrosion in NaCl solution (a) and fracture area (b) of the specimen corroded 288 h at 290°C.

CONCLUSIONS

Two kinds of tests, both under static and *quasi*-dynamic conditions, were applied in order to understand dissolution mechanism of Si_3N_4 and sialon ceramics under subcritical conditions in neutral aqueous solutions. The tests carried out under static conditions enabled identification of precipitated corrosion products after saturation was achieved under the conditions of most severe attack at 290°C. The corrosion tests performed under *quasi*-dynamic conditions facilitated determination of initial dissolution rates at 290°C and calculations of activation energies. Both studied ceramics dissolved by preferential attack of Si-N bonds in the matrix accompanied by the release of ammonia and formation of protective, mostly oxide layer of corrosion products at the surface. The original assumption that the corrosion resistance of Si_3N_4-based materials can be enhanced by dissolution of alumina in the crystal structure, creating sialon in the process, was not confirmed. The likely mechanism of corrosion of sialon ceramics was preferential disruption of Si-N and Al-N bonds in (Si, Al)ON₂ tetrahedra, with aluminium released into solution precipitating in the form of insoluble hydroxyaluminosilicates once the saturation concentration was attained. Penetration of aggressive chloride anions through protective layer of corrosion products is proposed to contribute to destruction of the layer and renewal of the exposure of vulnerable material surface to corrosion medium.

Considering the applications discussed in introduction of this paper the influence of corrosion under studied conditions on the bending strength of both selected ceramics are summarized in the Table III.

Table III. The studied ceramic materials sorting with respect to bending strength demand.

	Bending strength							
	De-ionized water				0.5 M NaCl			
	≤200°C 200 MPa	≤200°C 300 MPa	≤200°C 500 MPa	≤200°C 600 MPa	≤200°C 200 MPa	≤200°C 300 MPa	≤200°C 500 MPa	≤200°C 600 MPa
Si₃N₄	Y	Y	Y	n	Y	Y	Y	n
sialon	Y	n	n	n	/	/	/	/
	290°C / 200 MPa		290°C / 300 MPa		290°C / 200 MPa		290°C / 300 MPa	
Si₃N₄	Y		Y		Y		n	
sialon	Y		Y		Y		n	

Y - suitable for application with respect to bending strength
n - not suitable for application with respect to bending strength
/ - tests not performed for ≤200°C in NaCl solution

Considering obtained information based on experimental results all studied materials do appear to be suitable for application operating in aqueous sodium chloride solution at the temperatures ≤290°C and where requirements for bending strength did not exceed 200 MPa. High corrosion resistance of these studied ceramic materials at temperatures up to 200°C predestines them to be suitable candidates, e.g. in a form of coating, for increasing life-performance of exhaust pipes in aqueous sodium chloride environment.

ACKNOWLEDGMENT
This publication was created in the frame of the project "Centre of excellence for ceramics, glass, and silicate materials" ITMS code 262 201 20056, based on the Operational Program Research and Development funded from the European Fund of Regional Development. The financial support of this work by the APVV grant No APVV 0171-06 and by the grant of the Slovak Scientific Grant Agency VEGA under the contract number VEGA 2/0036/10 is gratefully acknowledged.

REFERENCES
[1] H. Weingärtner, E. U. Franck, Supercritical water as a Solvent, *Agew. Chem. Int. Ed., Wiley-VCH Verlag GmbH&Co. KGaA, Weinheim*, **44**, 2672-2692, (2005).
[2] P.A. Marrone, G. T. Hong, Corrosion control methods in supercritical water oxidation and gasification processes, *J. of Supercritical Fluids*, **82**, 83-103 (2009).
[3] P. Kritzer, N. Boukis, E. Dinjus, Factors controlling corrosion in high-temperature aqueous solutions: a contribution to the dissociation and solubility data influencing corrosion processes, *J. if Supercritical Fluids*, **15**, 205-227, (1999).
[4] P. Kritzer, Corrosion of high-temperature and supercritical water and aqueous solutions: a review, *J. of Supercritical Fluids*, **29**, 1-29, (2004).
[5] T. Laitinen, Localized corrosion of stainless steel in chloride, sulphate and thiosulfate containing environments, *Corrosion Science*, **42**, 421-441 (2000).
[6] I. Betova et al., Surface film electrochemistry of austenitic stainless steel and its main constituents in supercritical water, *J. of Supercritical Fluids*, **43**, 333-340 (2007).
[7] K.G. Nickel, U. Däumling and K. Weisskopf, Hydrothermal reactions of Si₃N₄, *Key Engineering Materials*, **89-91**, 295-300, (1994).
[8] M. Yoshimura, J.I. Kase, M. Hayakawa, S. Somiya, Oxidation Mechanism of Nitride and Carbide Powders by High-Temperature, High-Pressure Water, *Ceramic Transactions*, **10**, (1990).
[9] K. Oda, T. Yoshio, Y. Miyamoto, M. Koizumi, `Hydrothermal Corrosion of Pure, Hot Isostatically Pressed Silicon Nitride`, *J. Am. Ceram. Soc.*, **76**, 1365-1368, (1993).

[10]S. Somiya, `Hydrothermal corrosion of nitride and carbide of silicon`, *Mater. Chemistry and Physics*, **67**, 157-164, (2001).

[11]N. S. Jacobson, E. J. Opila, K. N. Lee, Oxidation and corrosion of ceramics and ceramic matrix composites, *Current Opinion in Solid State and Materials Science*, **5**, 301-309 (2001).

[12]M. V. Fedotova, Structural features of concentrated aqueous NaCl solution in the sub- and supercritical state at different densities, *J. of Molecular Liquids*, **143**, 35-41, (2008).

[13] A. Helebrant, Kinetics of Corrosion of Silicate Glasses in Aqueous Solutions, *Ceram.-Silik.*, **41** (4), 147-151, (1997).

[14]D. Galuskova, Korózia konštrukčných keramických materiálov vo vodnom roztoku chloridu sodného, *PhD thesis*, Slovak Technical University, Faculty of Chemical and Food Technology, Bratislava, (2010).

[15]J. Schilm, M. Herrmann, G. Michael, Corrosion of Si$_3$N$_4$-ceramics in aqueous solutions. Part II: Corrosion mechanisms in acids as a function of concentration, temperature and composition, *J. Eur. Ceram. Soc.*, **27**, 3573-3588, (2007).

[16]T. Sato, Y. Tokunaga, T. Endo, M. Shimada, Corrosion of Silicon Nitride Ceramics in Aqueous Chloride Solutions, *J. Am. Ceram. Soc.*, **71** [12], 1074-79, (1988).

[17]D.L.Parkhust, C.A.J. Appelo, User's Guide to PHREEQC (Version 2). http://wwwbrr.cr.usgs.gov/projests/GWC_coupled/phreeqc/index.html, U.S. Geological Survey, Denver, CO, Water-Resources Investigations Report 99-4259 (1999).

SMART RECYCLING OF COMPOSITE MATERIALS

Makio Naito, Hiroya Abe and Akira Kondo
Joining and Welding Research Institute, Osaka University
11-1, Mihogaoka, Ibaraki, Osaka, 567-0047, Japan

Masashi Miura and Norifumi Isu
INAX Corporation
3-77, Minatomachi, Tokoname, Aichi, 479-8588, Japan

Takahiro Ohmura
NICHIAS Corporation
1-8-1, Shin-miyakoda, Kita-ku, Hamamatsu, Shizuoka, 431-2103, Japan

ABSTRACT

Smart powder processing stands for novel powder processing techniques that create advanced materials with minimal energy consumption and environmental impacts. Particle bonding technology is a typical smart powder processing technique to make advanced composites. It creates direct bonding between particles without any heat support or binders of any kind in the dry phase. The bonding is achieved through the enhanced particle surface activation induced by mechanical energy, in addition to the intrinsic high surface reactivity of nanoparticles. Using this feature, desired composite particles can be successfully fabricated. It can also custom various kinds of nano/micro structures by using the composite particles, and can produce new materials with a simpler manufacturing process in comparison to wet chemical techniques.

By carefully controlling the bonding between different kinds of materials in the composite particles, effective separation of elemental components can be also achieved. It leads to the development of a novel technique for recycling advanced composite materials and turns them to high-functional applications. In this paper, this approach by using particle bonding and its disassembling will be introduced.

INTRODUCTION

Recently, various novel powder processing techniques were rapidly developed for advanced material production due to the growing of high-tech industry, especially in consideration of energy consumption and the environmental issues such as the recycling of waste materials. Smart powder processing stands for novel powder processing techniques that create advanced materials with minimal energy consumption and environmental impacts. Particle bonding technology is a typical smart powder processing technique to make advanced composites.[1-5] It creates direct bonding between particles without any heat support or binders of any kind in the dry phase. The bonding is achieved through the enhanced particle surface activation induced by mechanical energy, in addition to the intrinsic high surface reactivity of nanoparticles. Using this feature, desired composite particles can be successfully fabricated. This technology has also its ability to control the nano/micro structure of the assembled composite particles. As a result, it can custom various kinds of nano/micro structures and can produce new materials with a simpler manufacturing process in comparison to wet chemical techniques.

By carefully controlling the bonding between different kinds of materials, separation of composite structure into elemental components is also possible, which leads to the development of novel technique to recycling advanced composite materials and turns them to high-functional applications. In this paper, the development of novel method to recycle glass-fiber reinforced plastics (GFRP) would be introduced.

PRODUCTION OF COMPOSITE PARTICLES

So far, various kinds of composite particles have been produced by particle bonding process. The typical one is a core particle coated with fine guest particles. Bonding fine particles on the surface of core particle was already proposed. Fig.1 showed the particle bonding process[3] by Mechanofusion System.[4] The process is carried out in two steps. First, the surfaces of fine particles and core particles are mechanically activated, as a result, fine particles adhere onto the surfaces of core particles. Then, as the second step, fine particles and core particles interact to each other while fine particles also adhere onto the fine-particle layer on the surfaces of core particles. Therefore, by changing the kind of fine particle materials during the processing, we can easily make porous layered particles or multi-layered composite particles.

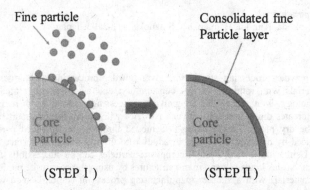

Fig.1. Particle composite processed by Mechanofusion System

There are a lot of factors affecting the particle composing process, and, its bonding mechanism depends on the combinations of core/fine particulate materials. However, as is well known, the contact surface between the powder materials receives extremely high local temperature and pressure, where mechanical stresses were actually given.[6] For example, authors previously reported that the local temperature at the interface between particles during particle bonding processing could be ten times higher than the apparent temperature of processing chamber.[7] Such a locally high temperature is expected to cause unique phenomena between fine particles and core particles, or among fine particles. For example, when coating nano titania particles on the glass beads, the peak of binding energy of Ti 2p shifted away from its original position after only 5 min of mechanical processing.[5] It suggested that there was a chemical interaction on their surface during the processing.

THE DEVELOPMENT OF NOVEL RECYCLING PROCESS FOR GFRP

By making use of particle bonding principle between different kinds of materials, disassembling them is also possible and it can be applied to recycle waste composite materials. For example, glass fiber reinforced plastics (GFRP) is a typical composite material having the advantages of lightweight, high strength and high weather resistance. Therefore, it has been used in various applications including boats, bath tubs, and building materials. Its production volume reached 460,000 tons in Japan in 1996, but decreased gradually since then. However, the volume of waste GFRP has increased every year. So far, almost all of the waste GFRP has been incinerated or disposed in landfill. Only 1-2 % of the waste GFRP is recycled as cement raw material or additives for concrete. Japan Reinforced Plastics Society started producing cement recycled from GFRP in 2002. The incineration of GFRP has problems of low calorific values on burning, and its residue needs to be disposed. In order to recycle the waste GFRP, some advanced chemical solvents and supercritical fluid [8] have been studied. However, they have not been used in practice, because the chemical approach requires high temperature and high pressure operating conditions, which are not only costly but also generate by-products. In addition, the recycled materials do not have similar quality to that of the starting materials.

GFRP usually contains 40-50% of calcium carbonate filler and 20-30% of glass fibers. These materials must be recycled through simple and low energy process for profits. Therefore, we aimed to develop new recycling method to make advanced materials from the waste GFRP. Fig.2 showed the concept of an innovative recycling process of GFRP proposed by the authors. [9]

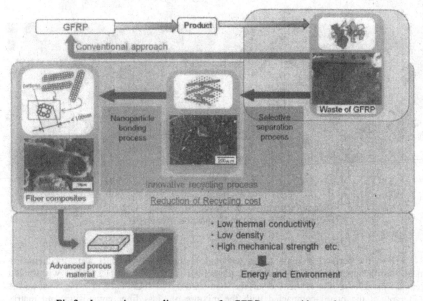

Fig.2. Innovative recycling process for GFRP proposed by authors

It consisted of two unit processes based on particle bonding principle. First, GFRP was separated into glass fibers and matrix resins, and then, the surface of separated glass fibers was coated by low cost nanoparticles with nanoparticle bonding process as shown in Fig.1. The coated composite glass fibers would be compacted to make porous materials. High functional materials having the properties of very low thermal conductivity because of the formation of nanoporous layer, light weight, and easy machining are expected to obtain by applying the new process shown in Fig.2.

The waste GFRP chip crushed down to about 1 cm was processed by an attrition-type mill, which applied similar mechanical principle to that of particle bonding process. When strong shear stress was applied to the chip layers for surface grinding, glass fibers began to separate from matrix resins on the chip surfaces. As a result, all glass fibers were effectively separated from other matrix components. The SEM photographs of the processed waste GFRP were shown in Fig.3.[9] It was obvious that the glass fibers separated from matrix resins had their own shape by using this method (Fig.3 (a)). The length of the glass fibers ranged from about 100 m to over 1 mm. On the other hand, the glass fibers were destroyed when applying an impaction milling (Fig.3 (b)). In the impaction milling, mainly impact force was applied to the material, which reduced glass fibers to particle form. These results showed that the proposed method using particle bonding principle was very effective in selective separation of glass fibers from other matrix components.

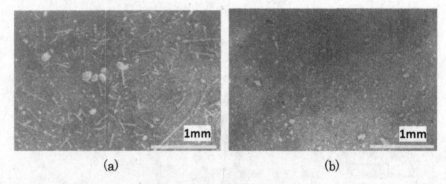

(a)   (b)

Fig.3. SEM photographs of processed powders by using the
proposed method (a) and impaction milling (b)

Then, the surface of glass fibers and that of matrix components were mechanically bonded with nanoparticles. Fig.4 shows the glass fiber composite particles coated with fumed silica nanoparticles. It shows that the surface is covered by nanoporous layer.[10] Fig.5 shows the photograph of the board compacted with the mixture of glass fiber composites and matrix components by dry pressing. The board has relatively high fracture strength, therefore it was easy for machining into various shapes.

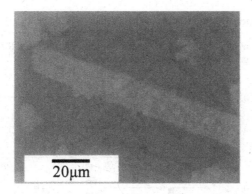

Fig.4. SEM image of glass fiber composite particles coated with fumed silica

Fig.5. Fiber reinforced porous fumed silica compact

Fig.6 shows the relationship between thermal conductivity and bulk density of the compact.[10] When fumed silica mass percentage increases, the bulk density decreases, thus leads to extremely lower thermal conductivity of the compact as shown in Fig.6. It means that the board made by this method can be used as high performance thermal insulation materials. Fig.7 shows the relationship between flexural strength and bulk density of the compacts. There are several processing methods to increase the fracture strength of the board. The typical method is to increase pressure by dry pressing. As shown in Fig.7, higher flexural strength is obtained by increasing the compressive pressure.[10] Another method is to increase cohesion force between particles in bulk compact.[11] This would be better not to increase the bulk density to obtain higher mechanical strength of the bulk compact. Further experiments will be needed to achieve higher flexural strength of the bulk compacts in future.

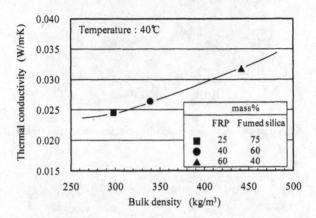

Fig.6. Relationship between thermal conductivity and bulk density of the compacts

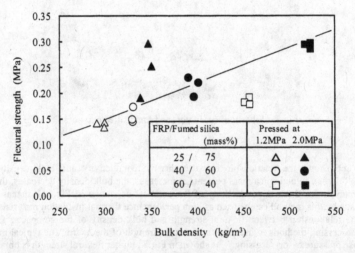

Fig.7. Relationship between flexural strength and bulk density of the compacts

CONCLUSION

In this paper, particle bonding process was explained as a typical example of smart powder processing. By applying the particle bonding process and its disassembling, new recycling process for waste GFRP and turn it into other advanced materials was developed.

ACKNOWLEDGMENT

This work was partially supported by Research & Technology Development on Waste Management Research Grant of Ministry of the Environment.

REFERENCES

[1] M.Naito and H.Abe, Particle Bonding Technology for Composite Materials– Microstructure Control and its Characterization, *Ceramic Transactions*, **157**, 69-76 (2004).

[2] M.Naito, H.Abe, and K.Sato, Nanoparticle Bonding Technology for the Structural Control of Particles and Materials, *J. Soc. Powder Technol., Japan*, **42**, 625-631 (2005).

[3] M.Naito, A.Kondo and T.Yokoyama, Applications of Comminution Techniques for the Surface Modification of Powder Materials, *ISIJ International*, **33**, 915-924 (1993).

[4] T.Yokoyama, K.Urayama, M.Naito, M.Kato and T.Yokoyama, The Angmill Mechanofusion System and its Applications, *KONA*, **5**, 59-68 (1987).

[5] M.Naito and M.Yoshikawa, Applications of Mechanofusion System for the Production of Superconductive Oxides, *KONA*, **7**, 119-122 (1987).

[6] F.Dachille and R.Roy, High-pressure Phase Transformations in Laboratory Mechanical Mixers and Mortars, *Nature*, **186**, 34 (1960).

[7] K.Nogi, M.Naito, A.Kondo, A.Nakahira, K.Niihara and T.Yokoyama, New Method for Elucidation of Temperature at the Interface between Particles under Mechanical Stirring, *J. Japan Soc. Powder and Powder Metall.*, **43**, 396-401 (1996).

[8] T.Iwaya, S.Tokano, M.Sasaki, M.Goto and K.Shibata, Recycling of Fiber Reinforced Plastics Using Depolymerization by Solvothermal Reaction with Catalyst , *J. Mater. Sci.*, **43**, 2452-2456 (2008).

[9] M.Naito, H.Abe, A.Kondo, T.Yokoyama and C.C.Huang, Smart Powder Processing for Advanced Materials, *KONA Powder and Particle Journal*, No.27, 130-141 (2009).

[10] A.Kondo, H.Abe, N.Isu, M.Miura, A.Mori, T.Ohmura and M.Naito, Development of Light Weight Materials with Low Thermal Conductivity by Making Use of a Waste FRP, *J. Soc. Powder Technol., Japan*, **47**, 768-772 (2010).

[11] A.Kondo, H.Abe, M.Naito, Y.Okuni, M.Miura and N.Isu, Novel Recycling Process of Waste FRP for Advanced Materials, *Ceramic Transactions*, **219**, 229-235 (2010).

HIGH THERMAL CONDUCTIVITY AND HIGH STRENGTH SINTERED REACTION - BONDED SILICON NITRIDE CERAMICS FABRICATED BY USING LOW GRADE Si POWDER

Dai Kusano[a,b], Shigeru Adachi[b], Gen Tanabe[b], Hideki Hyuga[a], You Zhou[a], Kiyoshi Hirao[a*]
[a]National Institute of Advanced Industrial Science and Technology (AIST), Shimo-Shidami, Moriyama-ku, Nagoya 463-8560, Japan
[b]Japan Fine Ceramics Co., Ltd., Akedori, Sendai 981-3203, Japan

ABSTRACT

Sintered reaction bonded silicon nitrides (SRBSNs) were fabricated from commercially available low grade Si powder containing 1.6 mass % of impurity oxygen and 500 ppm of metallic impurities. Powder compacts of the raw Si powder doped with Y_2O_3 and $MgO/MgSiN_2$ as sintering additives were nitrided at 1400°C for 8h under a N_2 pressure of 0.1 MPa, followed by post-sintering at 1900°C for 6h under a N_2 pressure of 0.9 MPa. The SRBSN with Y_2O_3 and MgO as sintering additives had four-point bending strength of about 700 MPa, but lower thermal conductivity of about 89 W/m/K. Thermal conductivity could be improved to over 100 W/m/K without degrading bending strength by replacing MgO with $MgSiN_2$. It is thought that $MgSiN_2$ decreased the amount of oxygen in the liquid phase formed during sintering which resulted in reduced lattice oxygen content in the developed β-Si_3N_4 grains. These good properties were equivalent to those of the SRBSN fabricated by using a reagent grade high purity Si powder containing only 0.3 mass% of impurity oxygen.

INTRODUCTION

Power supply and packing density of power modules are rapidly increasing as their application field expands, in particular in the automobile industry. In order to guarantee the stable operation of the power module, heat release technology in the module becomes very important. So far AlN has been used as a substrate for power devices because it exhibits high thermal conductivity over 200 W/m/K [1]. In general, a copper plate as an electrode is directly bonded to a ceramic substrate via high temperature heat treatment, which causes large residual thermal stress in the substrate. The mechanical properties of AlN are not sufficiently high, which results in low reliability of substrates. The electronic industry, therefore, is eager to seek alternative substrate materials.

Silicon nitride is well known as a typical structural ceramic which exhibits many excellent properties such as high strength, high toughness and high refractoriness. In addition to these properties, β-Si_3N_4 single crystal itself has been proven to possess a high intrinsic thermal conductivity over 200 W/m/K [2]. However, the thermal conductivity of sintered silicon nitride is generally much lower than the predicted value of high purity β-Si_3N_4 single crystal [5]. That is because imperfections in β-Si_3N_4 grains such as point defects as well as grain boundaries and secondary phases with low thermal conductivity negatively affect the thermal conductivity of sintered Si_3N_4 [3].

Recently, a research group at AIST has succeeded in fabricating Si_3N_4 with both high thermal conductivity and high strength via a reaction bonding followed by post sintering process[9, 10]. For instance, Zhou et al. fabricated sintered reaction-bonded silicon nitride (SRBSN) materials by nitriding a high-purity Si powder compact doped with Y_2O_3 and MgO as sintering additives followed by post-sintering (>100W/m/K, >700MPa)[10].

The above-mentioned experiments have been conducted using high purity silicon powders to demonstrate the potential of reaction bonding process. From the industrial view points, various kinds of silicon powders including those with low purity and/or coarse particle size should be employed as raw powders. In the present study coarse silicon powder with low purity was used for fabricating SRBSN, and the thermal and mechanical properties of the SRBSN were evaluated and compared with those of a SRBSN prepared from a high purity Si powder.

EXPERIMENTAL PROCEDURE

In this study commercially available Si powder with low-purity (purity > 99.9%) was used as a starting raw powder (hereafter, referred to as powder L). For the sake of comparison reagent grade high purity Si powder (purity > 99.99%) was also used (hereafter, referred to as powder H). For both Si powders BET specific surface (AUTOSORB-3B, Yuasa-Ionics Co., Ltd., Osaka, Japan) and impurity oxygen content (Model TC-436, LECO Co., St. Joseph, MI) were measured. The characteristics of these powders are summarized in Table 1. Powder H contains a very small amount of oxygen (0.28 mass%), and its metallic impurity is negligible; while powder L contains a large amount of oxygen impurity (1.58 mass%) with some metallic impurities. The mean particle sizes of powders L and H measured by a particle size analyzer (Model HRA9320-X100, Nikkiso Co., Tokyo, Japan) were 5.27 and 8.50 µm, respectively. The morphology of the powders was observed by scanning electron microscopy (SEM, Model JSM-5600, JEOL Ltd., Tokyo, Japan).

Table 1. Characteristics of the two Si raw powders

Sample		Powder L	Powder H
Purity	[%]	>99.9	>99.99
d_{50} (as received)	[µm]	5.27	8.50
BET (as received)	[m²/g]	3.21	1.70
BET (after milling)		6.76	4.83
Oxygen content (as received)	[mass%]	1.58	0.28
Impurity	[%]	Fe : 0.019	0.0011
		Ca : 0.015	<0.0001
		Mn : ?	<0.0001
		Al : 0.018	0.0002

Y_2O_3 (purity> 99.9%, Shin-Etsu Chemical Co. Ltd., Tokyo, Japan) and MgO (purity >99.9%, UBE Industries Ltd., Yamaguchi, Japan) were added as sintering additives to the two Si raw powders of L and H (referred to as specimen L-YMO and H-YMO, respectively). $MgSiN_2$ (made by ourself [11]) was also employed as a magnesium compound in the experiment using powder L (referred to as specimen L-YMN). Molar ratios of the starting compositions were Si_3N_4 : Y_2O_3 : MgO = 93 : 2 : 5 and Si_3N_4 : Y_2O_3 : $MgSiN_2$ = 93 : 2 : 5, respectively, on the assumption that Si powder is completely nitrided to Si_3N_4. The Si powder and sintering additives were mixed in ethanol using a planetary mill in a Si_3N_4 jar with Si_3N_4 balls. The rotation speed was 250 rpm and the grinding time was 2 h. After the vacuum drying and sieving, about 18 g of the mixed powder was pressed into a rectangular shape with 45 × 45 × 5 mm size using a stainless-steel die, followed by cold-isostatic pressing at 300 MPa. The green densities of the compacts were 1.40, 1.44, and 1.50 g/cm^3 for the specimens L-YMO, L-YMN and H-YMO respectively. The green compacts were placed into a BN crucible filled with BN powder, which was then set in a graphite crucible. The green compacts were nitrided at 1400 ℃ for 8 h under a nitrogen pressure of 0.1 MPa in a graphite resistance furnace (Multi-500, Fujidempa Kogyo Co. Ltd., Osaka, Japan). After nitridation, the nitrided compacts were sintered at 1900 ℃ for 6 h under a nitrogen pressure of 0.9 MPa in a graphite resistance furnace.

The microstructure of the nitrided specimens were observed using scanning electron microscope (SEM, JSM-5600, JEOL, Tokyo, Japan) and their phase identification was performed using X-ray diffraction analysis (XRD, RINT2500, Rigaku Corporation, Tokyo, Japan) with CuKα radiation of 40 kV / 100 mA. In addition, the quantitative analysis of α- and β-Si_3N_4 phases was also carried out according to the method reported by Pigeon and Varma[12].

The densities of SRBSN were measured using Archimedes' method with distilled water as an immersion medium. Then, the specimens were cut into 3 ×4 × 36 mm bars, after that their surfaces were ground with a 400-grit diamond wheel prior to the flexure tests. Mechanical properties were evaluated by four-point bending test. The microstructures of fracture surfaces were observed using SEM, and phase compositions were identified using XRD. Thermal diffusivity was measured by the laser flash method[10, 15] (Model TC-7000, ULVAC, Yokohama, Japan). Thermal conductivity (κ) was calculated by the following equation:

$$\kappa = \rho C_p \alpha$$

where ρ, C_p, and α are the bulk density, specific heat, and thermal diffusivity, respectively. The specific heat values of the SRBSNs fabricated in this work were very similar. Therefore, a constant value of specific heat, 0.68 J/g/K, was used to calculate thermal conductivity for all the SRBSNs in this work [8].

RESULT AND DISCUSSION

(1) Characteristics of raw Si powders

At first, characteristics of the two Si powders before and after milling were examined. The SEM micrographs and particle size distributions of the two as-received Si powders (powders L and H) are shown in Figure 1. Powder L was composed of coarse particles with diameter of about a few tens of microns and fine particles with diameter of about a few microns (Fig. 1-a), and exhibited a broad particle size distribution as shown in Fig. 1-(c). On the contrary, coarse particles with diameter of a few tens of microns dominated in powder H (Figs. 1-b), and it exhibited a rather sharp size distribution as shown in Fig. 1-d. Reflecting such particle distributions, specific surface area for the powder L is higher than that for the powder H as shown in Table 1.

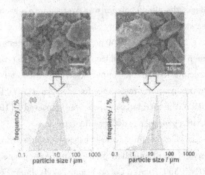

Fig. 1 SEM micrographs and particle distributions of the two as-received Si powders (Powder H and Powder L).

Figure 2 shows the SEM micrographs and particle size distributions of the two Si powders after milling for 2 h. It seems that larger particles in powder L were pulverized to be less than 10 microns, with the sizes of most particles ranging from 0.2 to 10 m (Fig. 2-c). Powder H was also pulverized in a similar manner.

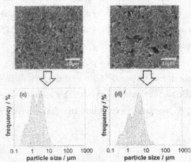

Fig. 2 SEM micrographs and particle distributions of the two ground Si powders (Powder H and Powder L).

In both Si powders oxygen content increased with increasing milling time as shown in Fig. 3. After 2 h milling the oxygen contents of powder L and H increased from 1.58 to 1.81 mass%, and from 0.28 to 0.51 mass%, respectively. The increased amount of oxygen in both cases was about the same (nearly 0.2 mass%). It is thought that the increase in oxygen content was attributed to the oxidation of the newly formed Si surfaces by fracture of particles. In general, surface of each Si particle is covered with a thin oxide layer [13], therefore the impurity oxygen in the high-purity Si powder was expected to mainly exist as the surface oxide layer. In this sense it is well understood that in powder H the oxygen content after 2 h milling doubled from the as-received, being associated with about 2.8 times increase in the specific surface area. That is, the oxygen content increased from 0.28 mass% to 0.51 mass% (the circles in Fig. 3), and the specific surface area increased from 1.70 to 4.83 m²/g (Table 1). However, concerning powder L, even in the as-received state the amount of oxygen impurity was 1.58 mass% which was much higher than that of the milled H powder, although its specific surface area of 3.21 m²/g was smaller than the milled powder H. It is likely that the oxygen in powder L existed as silica and metal oxides.

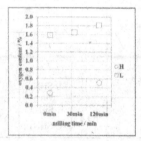

Fig.3 Relationship between milling time and oxygen content.

(2) Nitridation of Si powder compacts

The powder compacts of L-YMO, L-YMN and H-YMO compositions were nitrided at 1400 ℃ for 8 h. The relative densities of the three nitrided compacts were different: 70.7% for L-YMO, 69.8% for L-YMN, and 74.5% for H-YMO, respectively. Nitriding ratios calculated based on weight gain were at the same level for the three nitrided specimens. It has been reported that nitriding ratio measured from weigh gain was around 95% due to evaporation of Si during heating [10].

Phase identification of the nitrided compacts was conducted, and the results are summarized in Table 2. For all the nitrided compacts residual Si was not detected, indicating complete nitridation of Si powder. The major phases in all the compacts were α and β Si_3N_4. When focusing attention on the secondary phases, $YSiO_2N$ and Y_2SiO_5 were observed in the L-YMO and L-YMN compacts, while only $YSiO_2N$ was observed in the H-YMO compact. These secondary phases were formed by the reaction of

Si_3N_4, the Y_2O_3 additive, and impurity SiO_2 in the Si raw powder.

The phase diagram of $Si_3N_4 \cdot Y_2O_3 \cdot SiO_2$ system is shown in Fig. 4. From the diagram the equilibrium phases in the vicinity of Si_3N_4 corner is shifted in the sequence of $Si_3N_4 \cdot YSiO_2N$ $\cdot Y_2Si_5N_4O_3$, $Si_3N_4 \cdot YSiO_2N - Y_2SiO_5$, $Si_3N_4 \cdot Y_2SiO_5 \cdot Y_2Si_2O_7$, and $Si_3N_4 \cdot Y_2Si_2O_7 - Si_2N_2O$. Therefore, it is quite reasonable that the Y_2SiO_5 phase was detected in the nitrided specimens from the powder L with larger amount of impurity oxygen. In addition, the peak intensity of Y_2SiO_5 was lower in the L-YMN compact compared to the L-YMO, which may be related to the decrease in the total oxygen in the mixed powder by replacing MgO with $MgSiN_2$. Phases containing magnesium element could not be detected. The magnesium is thought to exist as the glassy phase in the Si-Y-Mg-O-N system.

Table 2 The result of the nitridation of L-YMO, L-YMN and H-YMO under a N_2 pressure of 0.1 MPa at 1400 ℃ for 8 h

	Nitrided compact				
	Nitridation [%]	R.D. [%]	XRD		
			+++	++	+
L-YMO	95.3	70.7	$\alpha\text{-}Si_3N_4$, $\beta\text{-}Si_3N_4$	$YSiO_2N$, Y_2SiO_5	—
L-YMN	93.7	69.8	$\alpha\text{-}Si_3N_4$, $\beta\text{-}Si_3N_4$	$YSiO_2N$	Y_2SiO_5
H-YMO	95.5	74.5	$\alpha\text{-}Si_3N_4$, $\beta\text{-}Si_3N_4$	$YSiO_2N$	—

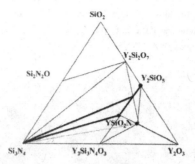

Fig. 4 Phase diagram of Si₃N₄ – Y₂O₃ – SiO₂[16].

(3) Post-sintering of reaction-bonded silicon nitrides

The characteristics of the SRBSN are shown in Table 3. All of the specimens could be densified to nearly full density.

Table 3 The characteristics of the sintered materials obtained by post-sintering under a N_2 pressure of 0.9 MPa at 1900 °C for 6 h

	Sintered body			
	Relative Density [%]	Weight Loss [%]	Bending strength [MPa]	Thermal conductivity [W/m/K]
L-YMO	99.1	3.91	727.6	88.8
L-YMN	98.9	2.69	716.0	100.3
H-YMO	99.5	1.71	716.3	101.2

The SEM micrographs of the SRBSNs are shown in Fig.5. All of the specimens were basically made up of rodlike elongated grains, but difference in the microstructures of the three materials could be seen. The L-YMN and H-YMO specimens exhibited distinct bimodal microstructures with large elongated grains, several tens of microns in length and a few microns in width, being distributed in smaller grains matrix; while the grain size distribution was narrow and a large amount of fine roundish grains could be seen in the L-YMO specimen. It seems that grain growth in the L-YMO was suppressed compared to the L-YMN and H-YMO specimens.

Tajika et al. [14] reported that, in the liquid phase sintering of AlN, the grain growth of AlN tended to become slow with increasing amount of liquid phase. Because the grain growth rate was decided by the diffusion of AlN through the boundary phases. It is thought that grain growth of β-Si₃N₄ was constrained by larger amount of liquid phase in the L-YMO specimen. On the contrary, in the L-YMN, for which MgSiN₂ was employed as one of the sintering additives, grain growth might have been enhanced in the nitrogen rich liquid phase.

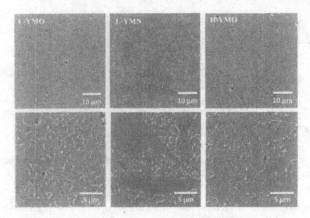

Fig. 5 SEM micrographs of the sintered bodies of L-YMO, L-YMN and H-YMO.

All of the SRBSN specimens had about the same bending strength as shown in Table 3. Thermal conductivity of the L-YMO specimen was as low as 88.8 W/m/K, but it could be improved to over 100 W/m/K by using MgSiN₂ as one of the sintering additives in the L-YMN specimen. It should be noticed that thermal conductivity of the L-YMN was almost the same as that of the H-YMO specimen which was fabricated form the high purity Si powder.

Two reasons could be considered for the improved thermal conductivity of L-YMN specimen. One is a decrease in the amount of secondary phases. The other one might be decreasing of lattice oxygen resulted from the variation of liquid phase composition during sintering. It is thought that replacement of MgO with MgSiN₂ as a sintering additive might have resulted in a decrease in the amount of oxygen in the liquid phase at the sintering temperature, leading to the formation of β-Si₃N₄ with lower lattice oxygen during the solution · reprecipitation process[6,7]. Indeed, XRD analysis of the nitrided compacts supported this consideration. The secondary phases appeared in the L-YMO compacts shifted from SiO₂ less region in the L-YMN compact, as shown in Table 2.

SUMMARY

Commercially available Si powder containing a large amount of oxygen as well as metallic impurities was employed for fabricating SRBSN using Y₂O₃ and MgO/MgSiN₂ as sintering additives. The SRBSNs were fabricated by nitriding the green compacts at 1400 ℃ for 8 h under a N₂ pressure of 0.1 MPa, followed by post-sintering at 1900 °C for 6 h under a N_2 pressure of 0.9 MPa. The SRBSN with Y_2O_3 and MgO as sintering additives had a modest bending strength of about 700 MPa, but lower thermal conductivity of about 89 W/m/K. Thermal conductivity could be improved to over 100 W/m/K without degrading bending strength by replacing MgO with $MgSiN_2$. These good properties were equivalent to those of the SRBSN fabricated by using a reagent grade high purity Si powder.

REFERENCES

1. G. A. Slack, "Nonmetallic crystals with high thermal conductivity," J. Phys .Chem. Solids., 34 321 (1973).
2. J. S. Haggerty and A. Lightfoot, "Opportunities for Enhancing the Thermal Conductivity of SiC and Si_3N_4 Ceramics through Improved Processing," Ceram. Eng. Sci. Proc., 16 457-87 (1995).
3. M. Kitayama, K. Hirao, A. Tsuge, K. Watari, M. Toriyama and S. Kanzaki, "Thermal Conductivity of β-Si_3N_4: I , Effects of Various Microstructural Factors," J. Am. Ceram. Soc., 82 [11] 3105-3112 (1999).
4. H. Hayashi, K. Hirao, M. Kitayama, Y. Ymauchi, S. Kanzaki, "Effect of Oxygen Content on Thermal Conductivity of Sintered Silicon Nitride," J. Ceram. Soc. Jpn., 109 [12], 1046-50 (2001).
5. K. Hirao, K. Watari, H.Hayashi and M. Kitayama, "High Thermal Conductivity Silicon Nitride, " MRS Bull., 26 [6], 451-5 (2001).
6. M. Kitayama, K. Hirao, A. Tsuge, K. Watari, M. Toriyama and S. Kanzaki, "Thermal conductivity of β-Si_3N_4: II .Effect of Lattice Oxygen," J. Am. Ceram. Soc., 83 [8], 1985-92 (2000).
7. M. Kitayama, K. Hirao, A. Tsuge, M. Toriyama, and S. Kanzaki, "Oxygen Content in β-Si_3N_4 Crystal Lattice," J.Am. Ceram. Soc., 82 [11], 3263-65 (1999).
8. H. Hayashi, K. Hirao, M. Toriyama, S. Kanzaki, K. Itatani, "$MgSiN_2$ Addition as a Means of Increasing the Thermal Conductivity of β-Silicon Nitride," J. Am. Ceram. Soc., 84 [12], 3060-2 (2001).
9. X. W. Zhu, Y. Zhou, and K. Hirao, "Effect of Sintering Additive Composition on the Processing and Thermal Conductivity of Sintered Reaction- Bonded Si_3N_4," J. Am. Ceram. Soc., 87 [7], 1398–1400 (2004).
10. Y Zhou, X. Zhu, K. Hirao and Z. Lences, "Sintered Reaction-Bonded Silicon Nitride with High Thermal Conductivity and High Strength," Int. J. Appl. Ceram. Technol., 5 [2], 119-126 (2008).
11. Z. Lences, K. Hirao, Y. Yamauchi, and S. Kanzaki, "Reaction Synthesis of Magnesium Silicon Nitride Powder," J. Am. Ceram. Soc., 86 [7], 1088-93 (2003).
12. R. G. Pigeon and A. Varma, "Quantitative Phase Analysis of Si_3N_4 by X-Ray Diffraction," J. Mater. Sci. Lett., 11, 1370–1372 (1992).
13. B. T. Lee, J. H. Yoo, H. D. Kim, "Size effect of raw Si powder on microstructures and mechanical properties of RBSN and GPSed-RBSN bodies," Mater. Sci. Eng. A333, 306-313 (2002).
14. M. Tajika, H. Matsubara and W. Rafaniello, "Experimental and Computational Study of Grain

Growth in AlN Based Ceramics," J. Ceram. Soc. Japan, 105 [11], 928-933 (1997).

15. Parker. W. J, Jenkins. R. J, Butler. C. P, Abbott. G. L, "Flash Method of Determining Thermal Diffusivity, Heat Capacity, and Thermal Conductivity," J. Appl. Physics. 32, 1679-1684 (1961).

16. Mihael. K. C, G. Thomas and S. M. Johnson, "Grain-Boundary-Phase Crystallizaiton and Strength of Silicon Nitride Sintered with a YSiAlON Glass," J. Am. Ceram. Soc., 73 [6] 1606-12 (1990).

ELECTRICAL CONDUCTIVE CNT-DISPERSED Si_3N_4 CERAMICS WITH DOUBLE PERCOLATION STRUCTURE

Sara Yoshio, Junichi Tatami, Toru Wakihara, Tomohiro Yamakawa,
Katsutoshi Komeya, and Takeshi Meguro
Yokohama National University, Yokohama, Japan

ABSTRACT

The purpose of this study is to control the electrical conductivity of carbon nanotube- (CNT) dispersed silicon nitride (Si_3N_4) ceramics by double percolation. We prepared two kinds of CNT-dispersed Si_3N_4 ceramics. Samples with double percolation structures were prepared using mixed granules having various quantities of CNTs (0.7~2.6 vol%) and without CNTs in the proportion of 50 to 50. The powder mixtures were sintered using a spark plasma sintering technique to obtain fully dense CNT-dispersed Si_3N_4 ceramics. For comparison, single percolation samples with various quantities of CNTs (0.7~2.6 vol%) were also similarly prepared. The relative density of all the sintered bodies was over 97%. The single percolation sample with 0.7 vol% CNT did not show electrical conductivity because the quantity of CNT was lower than the percolation threshold. On the other hand, the electrical conductivity of the double percolation sample with 0.7 vol% CNT was about five orders of magnitude greater than that of the 0.7 vol% single percolation sample. Consequently, it was shown that it is possible to control the electrical conductivity of CNT-dispersed Si_3N_4 by double percolation.

INTRODUCTION

Silicon nitride (Si_3N_4) has unique properties, such as excellent hardness, high strength, high corrosion resistance, and high thermal conductivity. Because of these properties, Si_3N_4 ceramics have been applied to the fabrication of structural components, cutting tools, and bearings[1)-3)]. One of the features of Si_3N_4 ceramics is electrical insulation; however, electrical conductivity is needed in some applications. Carbon nanotubes (CNTs) have been used as fillers with high electrical conductivity, high elastic modulus, high strength, and a high aspect ratio[4)-6)]. In particular, CNTs have unusual dimensions–a nanometer diameter and a high aspect ratio–resulting in a low percolation threshold for electrical conductivity. Therefore, CNTs are good candidates to provide electrical conductivity in Si_3N_4 ceramics. In the previous research on CNT-Si_3N_4 composites, it has been reported that density and strength decrease with the reaction of CNTs and Si_3N_4, with a large quantity of CNTs, and/or with agglomeration of CNTs[7)-8)]. We reported that the strength of the CNT-Si_3N_4 composite increases with lowering firing temperatures or with a lowering quantity of CNTs[9)-10)]. In order to improve the strength of CNT-dispersed Si_3N_4 ceramics, a decrease in the quantity of added CNTs is needed in the range of the formation of CNT connections. The percolation theory was used to discuss the relationship of CNTs to electrical conduction. In this theory, CNTs should be homogeneously dispersed in the

CNT-Si_3N_4 composite (we call this a single percolation structure, because it has only one percolation of CNTs.). If the quantity of CNTs is below the threshold, electrical conductivity does not occur.

We here propose a novel concept for the control of electrical conductivity in CNT-dispersed ceramics, namely double percolation, involving two different percolations. Actually, only one previous study on CNF-polymer composites has been reported[11]. In this paper, CNT-dispersed Si_3N_4 ceramics with a double percolation structure were fabricated and the relationship between electrical conductivity and the double percolation structure was investigated.

EXPERIMENT PROCEDURE

To form the double percolation structure, we made two kinds of granules: granules with CNT and granules without CNT. The granules with CNTs were prepared by bead milling CNTs. The granules without CNT were prepared by conventional wet ball milling of the raw powders. The single percolation structure was formed by using only granules with CNT.

Fine high-purity powders of Si_3N_4 (SN-E-10, Ube Co. Ltd., Japan), Y_2O_3 (RU, Shin-Etsu Chemical Co., Japan), Al_2O_3 (AKP-30, Sumitomo Chemical Co., Japan), AlN (F grade, Tokuyama Co., Japan), TiO_2 (R-11-P, Sakai Chemical Co., Japan), and HfO_2 (Kojundo Chemical Lab. Co., Japan) were used as raw materials. The multiwall carbon nanotubes used in this study (VGCFs, Showa Denko Co., Japan) had diameters of 60 nm and lengths of 6 μm. Ethanol was used as the solvent. Polyethylenimine (EPOMIN SP-103, Nippon Shokubai. Co., Japan) of molecular weight 250 was used as the cation dispersant.

CNTs of 0.10.15 wt% against the ethanol and polyethylenimine five times the amount of CNT additive amount were added to ethanol. First, the CNT-dispersed slurry was prepared by ultrasonic treatment (VCX600, Sonics & Materials, Inc.) for 20 min, and the bead milling (Star Mill, Ashizawa Finetech Co., Japan) was carried out at 3000 rpm for 2 h using 0.3 mm Al_2O_3 beads (SSA-999S, Nikkato Co., Japan). Next, the Si_3N_4, Y_2O_3, Al_2O_3, AlN, TiO_2, and HfO_2 powders were added to the CNT slurry with 2 wt% of the dispersant (Seruna E503, Chukyoyushi Co.). Ball milling was carried out at 110 rpm for 48 h using 5 mm sialon balls to mix the powders. The weight ratio of the balls to the powder was 4.65. Powder mixtures were obtained by evaporating the ethanol. Paraffin (4 wt%, melting point: 46–48 °C, Junsei Chemical Co., Japan) and Bis(2-ethyhexyl) phthalate (2 wt%, Wako Junyaku Co., Japan) were added as the binder and lubricant, respectively, to make granules by sieving the mixed powder using a nylon mesh with openings of 250 μm. Subsequently, two kinds of granules were dry blended by shaker mixer (TURBURA®-T2F, Willy A. Bachofen AG Maschinenfabrik) for 1 h.

Table 1 Mixture ratio of granules with CNT and/or without CNT.

Sample name	Mixture fraction
0.7C100	Granule with CNT0.7vol%×100%
1.3C50	Granule with CNT1.3vol%×50%+without CNT×50%
1.3C100	Granule with CNT1.3vol%×100%
2.6C50	Granule with CNT2.6vol%×50%+without CNT×50%
2.6C100	Granule with CNT2.6vol%×100%

Table 1 shows the mixing ratio of the two granules. For example, for the sample 2.6C50, 2.6 indicates the CNT content in the granule with CNT, and 50 does the granule content; the average CNT content in the sample is 1.3 vol%. Next, the organic binder was eliminated at 500 °C for 3 h at 4 L/min N$_2$ flow. After dewaxing, the mixture granules were fired at 1550~1570 °C for 1 min in 20 MPa N$_2$ using the spark plasma sintering technique (DR. SINTER SPS-1050, SPS Syntex Inc., Japan). Finally, the relative density of the samples was measured using the Archimedes method. The microstructure was observed using a scanning electron microscope (JSM-6390, JEOL, Japan). Electrical conductivity was measured using the four-terminal method.

RESULTS AND DISCUSSION

Table 2 Relative density and electrical conductivity of the sintered bodies.

Sample name	Relative density (%)	Electrical conductivity (S/m)
0.7C100	98.7	2.7×10^{-12}
1.3C50	98.8	3.4×10^{-7}
1.3C100	97.6	8.2
2.6C50	98.0	1.6
2.6C100	97.6	22.1

Table 2 shows the electrical conductivity of the sintered bodies. The relative densities of all the sintered bodies were over 97%. The single percolation sample of 0.7 vol% CNT did not show electrical conductivity because it had fewer CNTs than the percolation threshold. The single percolation samples having 1.3 and 2.6 vol% CNT did show electrical conductivity because the CNT amounts were larger than the percolation threshold. Figure 1 shows SEM images of the sintered bodies with a single percolation structure. When a sample was observed by SEM without deposition processing, the insulator phase part appears bright, because of the charge up, and the conductive phase part appears dark. Corresponding to the result in Table 2, 0.7C100 of the insulator appears bright by the charge up. Meanwhile, 1.3C100 and 2.6C100, which have electrical conductivity, appear dark; the surface asperity of the sintered bodies can be observed.

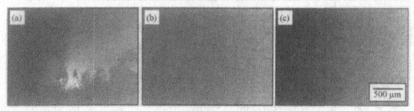

Figure 1 SEM images of the sintered bodies with single percolation structure. (a) 0.7C100, (b) 1.3C100, and (c) 2.6C100.

On the other hand, the electrical conductivity of the double percolation sample with 0.7 vol% CNT was about five orders of magnitude greater than that of the 0.7 vol% single percolation sample. However, the electrical conductivity of the double percolation sample with 1.3 vol% CNT was lower than that of the 1.3 vol% single percolation sample. Therefore, the percolation theory was used to analyze the samples. CNT-dispersed Si₃N₄ ceramics with double percolation structure have two scales of percolations: the first is percolation of the CNT, and the other is percolation of the granule with the CNT. The percolation of the granule with the CNT is represented by formula (1).

$$\sigma = \sigma_1 \left(f_{gra} - f_{c-gra} \right)^{v_{gra}}$$
(1)

In formula (1), σ is the electrical conductivity of the sample, f_{c-gra} is the percolation threshold of the conductive granule, v_{gra} is the percolation exponent, f_{gra} is the conductive granule content, and σ_1 is the exponent. In the case of 2.6C50, the electrical conductivity of the sample (σ) was calculated by substitution of the following values into formula (1): f_{c-gra}=0.3 (when assuming the simple cubic lattice), v_{gra}=1.3~2, f_{gra}=0.50, and σ_1=22.2 (electric conductivity of the conductive phase (2.6C100) obtained by this experiment). The result, σ=0.9~2.7 S/m, matched the experimental value (1.6 S/m). Therefore, it was shown that the electrical conductivity of the CNT-dispersed Si₃N₄ was controlled by double percolation. On the other hand, in the case of 1.3C50, the electrical conductivity of the sample (σ) was calculated by substitution of the following values into formula (1): f_{c-gra}=0.3 (when assuming the simple cubic lattice), v_{gra}=1.3~2, f_{gra}=0.50, and σ_1=8.2 (electric conductivity of the conductive phase (1.3C100) obtained by this experiment). The result is σ=0.3~1.0 S/m, and the experimental value (3.4×10^{-7} S/m) is smaller than the results (anticipated by the percolation theory). The sintered bodies with double percolation structures were then observed by SEM; the photograph is shown in Figure 2.

The photograph shows island-shaped bright insulators and the dark conductive phase in the sintered bodies; the conductive phases were connected to each other. In 2.6C50, the conductive areas were connected to each other in the conductive phase. However, most of the conductive areas were enclosed by the insulating area in the conductive phase in 1.3C50. This means that a conductive pass is not easily formed. Therefore, it is presumed that the electrical conductivity of 1.3C50 is smaller than the results anticipated by the percolation theory due to the minute insulator areas that exist between conductive phases.

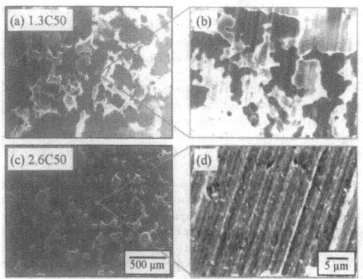

Figure 2 SEM images of CNT-dispersed Si₃N₄ ceramics with a double percolation structure. (b) and (d) are enlarged views of (a) and (c), respectively.

These results show that the electrical conductivity of CNT-dispersed Si₃N₄ can be controlled by double percolation.

CONCLUSION

CNT-dispersed Si₃N₄ ceramics with a double percolation structure were fabricated using granules with CNTs and subsequently evaluated. The single percolation sample with 0.5 wt% CNT did not show electrical conductivity because the quantity of CNT was less than the percolation threshold. On the other hand, the electrical conductivity of the double percolation sample having 0.5 wt% CNT was approximately five orders of magnitude greater than that of the 0.5 wt% single percolation sample. Consequently, it was shown that the electrical conductivity of CNT-dispersed Si₃N₄ can be controlled by double percolation.

REFERENCES

1H. Kawamura, and S. Yamamoto, Improvement of Diesel Engine Startability by Ceramic Glow Plug Start System, *Society of Automotive Engineers Paper* No. 830580 (1983).

2Y. Tajima, Development of High Performance Silicon Nitride Ceramics and Their Applications, *Mater. Res. Soc. Symp. Proc.*, **287**, 198–201 (1993).

3K. Komeya, and H. Kotani, Development of Ceramic Antifriction Bearing, *JSAE Rev.*, **7**, 72–9 (1986).

4S. Iijima, Helical Microtubules of Graphitic Carbon, *Nature*, **354**, 56 (1991).

5S. Rochie, Carbon Nanotubes: Exceptional Mechanical and Electrical Properties, *Ann. Chim. Sci. Mater.*, **25**, 529–32 (2000).

6H. J. Dai, Carbon Nanotubes: Opportunities and Challenges, *Surf. Sci.*, **500**, 218–41 (2002).

7Cs. Balázsi, Z. Kónya, F. Wéber, L. P. Biró, and P. Arató, Preparation and Characterization of Carbon Nanotube Reinforced Silicon Nitride Composites, *Mater. Sci. Eng.: C*, **23** [6–8] 1133–7 (2003).

8 C. Balazsi, B. Fenyi, N. Hegman, Z. Kover, F. Wéber, Z. Vertesy, Z. Kónya, I. Kiricsi, L.P. Biró, and P. Arató, Development of CNT/Si$_3$N$_4$ composites with improved mechanical and electrical properties, *Compos Part B-Eng.*, **37** [6] 418–24 (2006).

9J. Tatami, T. Katashima, K. Komeya, T. Meguro, and T. Wakihara, Electrically Conductive CNT-Dispersed Silicon Nitride Ceramics, *J. Am. Ceram. Soc.*, **88** [10] 2889–2893 (2005).

10S. Yoshio, J. Tatami, T. Wakihara, T. Yamakawa, H. Nakano, K. Komeya, and T. Meguro, Effect of CNT Quantity and Sintering Temperature on Electrical and Mechanical Properties of CNT-Dispersed Si$_3$N$_4$ Ceramics, *J Ceram Soc. Japan*, **119** [1], 70–75 (2011).

11C. Zhang, X.S. Yi, H. Yui, S. Asai, and M. Sumita, Selective location and double percolation of short carbon fiber filled polymer blends: high-density polyethylene/isotactic polypropylene, *Mater Let*, **36**, 186–190 (1998).

FABRICATION OF CNT-DISPERSED Si₃N₄ CERAMICS BY MECHANICAL DRY MIXING TECHNIQUE

Atsushi Hashimoto[1], Sara Yoshio[1], Junichi Tatami[1], Hiromi Nakano[2], Toru Wakihara[1], Katsutoshi Komeya[1], Takeshi Meguro[1]

1. Graduate School of Environment and Information Sciences, Yokohama National University
Yokohama 240-8501, Japan

2. Cooperative Research Facility Center, Toyohashi University of Technology
Toyohashi 441-8580, Japan

ABSTRACT

Carbon nanotube (CNT)- dispersed silicon nitride (Si_3N_4) ceramics were prepared with the mechanical dry mixing process by using CNT-TiO_2 nanocomposite particles. CNT-TiO_2 nanocomposite particles were prepared using a particle composer. TiO_2 nanoparticles were directly bonded on CNTs and the agglomerates of CNTs disappeared. CNT-TiO_2 nanocomposite particles were added to the silicon nitride and sintering aids by the wet mixing process. After firing the green bodies of the powder mixture by gas-pressure sintering technique, dense CNT-dispersed Si_3N_4 ceramics were obtained. The Si_3N_4 ceramics having 1.0 wt% CNTs exhibited electrical conductivity. The mechanical properties improved because the CNTs were homogenously dispersed.

INTRODUCTION

Silicon nitride (Si_3N_4) ceramics have excellent properties such as high strength, high toughness, high hardness, high corrosion resistance, high heat resistance, high thermal conductivity, and insulation. They have been applied to structural components such as bearings[1]. Si_3N_4 is an insulator; therefore, conventional Si_3N_4 ceramics are also insulators. Dust adheres to such Si_3N_4 ceramics owing to static electricity. As a result, adhesive dusts sometime cause breakdowns in the system and reduce the lifetime. To control the adhesive dusts, electrical conductivity is needed. In our previous study, the electrically conductive Si_3N_4 ceramics were fabricated by the wet mixing process[2]. Many CNTs remained in the Si_3N_4 grain boundary according to scanning electron microscope (SEM) observations. These CNTs form an electrical conductive path to induce electrical conductivity. However, there is a major issue with CNT-dispersed Si_3N_4 ceramics: CNTs frequently form agglomerates in Si_3N_4 ceramics fabricated by the wet mixing process, which become fracture origins. As a result, the agglomerates lower the strength of Si_3N_4 ceramics. Therefore, the dispersion of CNTs is a key process for improvement of the Si_3N_4 ceramics. There are several types of processes for the dispersion of CNTs, such as the chemical process and the mechanical process. However, in the chemical process, optimization of the dispersant and process conditions is needed. A wet bead milling process, which is one of the typical mechanical processes to disperse nanoparticles, is a very time-consuming process. Therefore, a simple and short process is desired for the homogenous dispersion of CNTs. Thus, we focused on the mechanical dry mixing technique. In this process, different types of powders are put in the chamber, and high shear stress is applied between particles by a high-speed rotating blade in a dry condition. Therefore, nanoparticles can be directly bonded with submicron particles and dispersed for a short time[3]. For example, we succeeded in preparing TiO$_2$-Si$_3$N$_4$ nanocomposite particles by this process and in fabricating the Si3N4 ceramics having excellent wear properties[4]. Therefore, this process may be applied to the dispersion of CNTs. The objectives of this study were to fabricate CNT-dispersed Si$_3$N$_4$ ceramics by the mechanical dry mixing technique and evaluate the electrical and mechanical properties of the developed CNT-dispersed Si$_3$N$_4$ ceramics.

EXPRIMENTAL PROCEDURE

The basic chemical composition of raw materials was Si$_3$N$_4$:Y$_2$O$_3$:Al$_2$O$_3$:AlN:TiO$_2$=92:5:3:5:5. The CNT quantity was changed from 0.5 wt% to 1.0 wt%. The CNTs were dispersed by dry mechanical mixing technique with TiO$_2$ under 1.5 kW for 10 min. After the mechanical dry mixing, particles were observed by a scanning electron microscope (SEM) and a transmission electron microscope (TEM). These powders were milled by wet ball milling and sieved to obtain granules using a 250-μm mesh sieve. The granules were molded to make green bodies. After dewaxing, they were fired at 1700-1800 °C in 0.9 MPa N$_2$ for 2 h followed by HIPping at 1700 °C in 100 MPa N$_2$ for 1 h. After machining the sintered compacts, the density was evaluated by Archimedes' method; the phase present was identified by X-ray diffraction (XRD); the microstructure was observed by SEM; the bending strength was measured by the three-point bending test; and electrical conductivity was measured by the DC 4-terminal technique.

RESULTS AND DISCUSSION

Figure 1 shows the SEM and TEM images of the microstructure after the mechanical dry mixing. It was observed that granules about 10 μm in diameter were formed, and there were no agglomerates of CNTs (Fig. 1 (a)). Figure 1(b) shows the SEM image after ultrasonication in ethanol for several minutes. The granules as shown in Fig. 1 (a) were easily disintegrated in ethanol. Moreover, TiO$_2$ and CNTs were well-dispersed and TiO$_2$ nanoparticles existed on the CNTs. Figure 1(c) shows the TEM image of the microstructure after mechanical dry mixing. It was confirmed that TiO$_2$ nanoparticles were bonded on the CNTs. Therefore, we succeeded in preparing TiO$_2$-CNT nanocomposites.

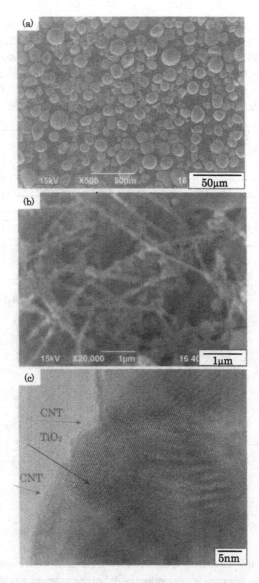

Figure 1. SEM images and TEM image of microstructure after mechanical dry mixing.
(a) After mechanical dry mixing, (b) after ultrasonication in ethanol and (c) TEM image

Table 1 lists the values of density, and Figure 2 shows the XRD pattern. The relative density of any specimen was 95% or more; therefore, we were able to obtain highly dense sintering compacts. Only β-Si₃N₄ and TiN were confirmed to have formed as a result of the identification of the phase present. The phases present were independent of the CNT quantity and the sintering temperature.

Table 1. Relative density, bending strength and electrical conductivity
of CNT-dispersed Si₃N₄ ceramics.

CNT quantity [wt%]	Firing temperature [°C]	Relative density after HIP [%]	3-point bending strength [MPa]	Electrical conductivity [S/m]
0.5	1700	95.6	705±39	—
	1750	94.9	945±70	—
	1800	95.0	820±87	—
1.0	1700	96.0	727±111	2.8
	1750	95.4	923±74	2.9
	1800	95.0	831±73	6.5

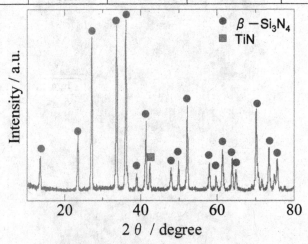

Figure 2. XRD pattern of CNT-dispersed Si₃N₄ ceramics.

Table 1 also lists the bending strength of CNT-dispersed Si₃N₄ ceramics. The 3-point bending strength of the samples fabricated by the mechanical dry mixing technique was higher than 700MPa. In particular, when the firing temperature was 1750 °C, the bending strength of CNT-dispersed Si₃N₄ ceramics was over 920MPa, which is almost the same value as the Si₃N₄ ceramics without CNTs. The fracture origin was observed after the 3-point bending test. The typical SEM image of the fracture origin after the 3-point bending test is shown in Figure 3. The circle shown in the figure represents the fracture origin. The fracture origin was not agglomerates of CNTs.

Figure 3. SEM image of fracture origin in the specimen with 1.0 wt% CNTs sintered at 1750°C

The electrical conductivity was also listed in Table 1. The specimen with 0.5 wt% CNTs did not exhibit electrical conductivity. On the other hand, the specimen with 1.0 wt% CNTs exhibited electrical conductivity. Figure 4 shows the SEM image of the fracture surface of the Si₃N₄ ceramics by adding 1.0 wt% CNTs fired at 1750 °C. Many CNTs remained, and there were no agglomerates of CNTs. These CNTs formed the electrical conductive path. In the specimen with 1.0 wt% CNTs, the electrical conductivity increased with the sintering temperature. According to our previous study[5], the electrical conductivity of CNT-dispersed Si₃N₄ ceramics increase with an increase in the grain size of -Si₃N₄. Generally, the grain size increases with the sintering temperature. In this study, the electrical conductivity increased with the sintering temperature. Therefore, a similar tendency may have been observed in this study.

Figure 4. SEM image of the fracture surface of the Si$_3$N$_4$ ceramics by adding 1.0 wt% CNTs fired at 1750 °C

CONCLUSIONS

CNT-dispersed Si$_3$N$_4$ ceramics were easily fabricated by the mechanical dry mixing technique. Samples with 1.0 wt% CNTs exhibited electrical conductivity. The mechanical properties improved because the CNTs were homogeneously dispersed.

REFERENCES

[1]K. Komeya and H. Kotani, "Development of Ceramic. Antifriction Bearing" *JSAE Rev.*, 7, 7279 (1986)

[2]J. Tatami, T. Katashima, K. Komeya, T Meguro and T. Wakihara, "Electrically Conductive CNT-Dispersed Silicon Nitride Ceramics" *J. Am. Ceram. Soc.*, 88 [10] 2889–2893 (2005)

[3]H. Abe, I. Abe, K. Sato and M. Naito, "Dry Powder Processing of Fibrous Fumed Silica Compacts for Thermal Insulation" *J. Am. Ceram. Soc.*, 88[5], 1359–1361 (2005)

[4]J. Tatami, E. Kodama, H. Watanabe H. Nakano, T. Wakihara, K. Komeya, T. Meguro and A. Azushima, "Fabrication and wear properties of TiN nanoparticle-dispersed Si$_3$N$_4$ ceramics" *J. Ceram. So. Japan.*, 116 [6] 749-754 (2008)

[5]S. Yoshio, J. Tatami, T. Wakihara, T. Yamakawa H. Nakano, K. Komeya and T. Meguro, "Effect of CNT quantity and sintering temperature on electrical and mechanical properties of CNT-dispersed Si$_3$N$_4$ ceramics" *J. Ceram. So. Japan.*, 119 [1] 70-75 (2011)

JOINING OF SILICON NITRIDE LONG PIPE BY LOCAL HEATING

Mikinori HOTTA, Naoki KONDO and Hideki KITA
Advanced Manufacturing Research Institute, National Institute of Advanced Industrial Science and Technology (AIST)
2266-98 Shimo-Shidami, Moriyama-ku, Nagoya 463-8560, Japan

ABSTRACT

Silicon nitride (Si_3N_4) pipes were joined to fabricate the long pipe of 2 meter in length by using local heating. The liquid-phase sintered Si_3N_4 ceramics with Y_2O_3 and Al_2O_3 additives were used as joining pipes. The powder mixture of Si_3N_4-Y_2O_3-Al_2O_3-SiO_2 system was inserted between the two pieces of Si_3N_4 pipe. To join the Si_3N_4 pipes, the joint region was selectively heated at 1600°C for 1 h at a mechanical pressure of 5 MPa in N_2 atmosphere using an electrical furnace. The Si_3N_4 long pipe of 2 meter in length was successfully fabricated by joining two Si_3N_4 pipes of 1 meter in length. Neither break nor void was observed at the joint region by optical microscopy. Flexural strength of the test specimens cut from the joined Si_3N_4 long pipe showed the average value of 680 MPa.

INTRODUCTION

Silicon nitride (Si_3N_4) ceramics are applied to industrial components in various manufacturing industries, for example, heating element protection tubes for handling molten aluminum, conveying rolls in steel production line and a variety of pipes, because of their excellent wear resistance, heat resistance, good corrosion resistance and lightweight property. The ceramic tubular components often require a large size over 1000 mm long. When the large components are produced as single unit, there are several problems; for examples it is necessary to use huge manufacturing facilities and it is difficult to perform a green machining prior to sintering. One of the techniques to overcome these problems is joining. In cases of long tubular components, it is necessary to join the components by using local heating around the joint region in order to lower consumed energy and production-cost. Furthermore, the Si_3N_4 ceramic components are used mostly under a high temperature and corrosion environment. In order to add the resistance to these environments, the joining layer should be composed of constituent and microstructure similar to the Si_3N_4 bulk in the joined Si_3N_4 ceramic components.

Many researchers have reported on the joining of various bulk ceramics including Si_3N_4[1-7]. However, the size of the bulk ceramics remained in test piece about 20 mm for bending strength. For the scale-up of Si_3N_4 tubular components, we recently joined block and pipe of larger size by using general heating and microwave local heating[8].

In the present work, the local heating equipment for joining of tubular ceramics was designed and developed. Si_3N_4 pipes were joined to fabricate Si_3N_4 long pipe of 2000 mm in length by using the local heating, toward a practical applications of Si_3N_4 ceramic tubular components.

EXPERIMENTAL PROCEDURE

Designing of Local Heating Equipment for Joining

The local heating equipment was designed to apply superplastic phenomenon of ceramics to joining of tubular ceramics. One side of the ceramic pipe is fixed in the equipment, and the joint region of the pipes is heated with an electric furnace to fabricate a ceramics long pipe. The heating zone at the furnace is 400 mm in length. The maximum size of joining pipe for the joining in the local heating equipment is 60 mm in outer diameter and 3000 mm in length. The fixture of the joining pipes in the outer side of the furnace enables to apply a mechanical pressure to the joint region in the pipes, up to 200 kgf.

Joining of Silicon Nitride Pipes

As bulk material, commercially available Si_3N_4 pipe sintered with 5 mass% Y_2O_3 and 5 mass% Al_2O_3 was used. The size of the Si_3N_4 pipes was 28 mm in outer diameter, 18 mm in inner diameter and 1000 mm in length. As an insert material, a powder mixture of Si_3N_4-Y_2O_3-Al_2O_3-SiO_2 in a composition of SiAlON glass was used. The composition of the mixed powder was 30.2 mass% Si_3N_4-43.3 mass% Y_2O_3-11.7 mass% Al_2O_3-14.8 mass% SiO_2, which was previously reported by Xie et al.[3]. A small amount of ethanol was added to the sieved mixture to prepare homogeneous slurry for the insert. The slurry was put on the joint surface of the Si_3N_4 pipes, and then dried. The joint region of the Si_3N_4 pipes was inserted into the furnace of the local heating equipment. The Si_3N_4 pipes were joined at a heating temperature of 1600°C for 1 h in flowing N_2 gas of 5 L/min at a heating rate of 10°C/min. The mechanical pressure of 5 MPa was applied to the joint surface during the joining of the Si_3N_4 pipes. The temperature of the Si_3N_4 pipe surface at the joint region was measured with an optical pyrometer.

Evaluation of Joined Silicon Nitride Pipe

The joined Si_3N_4 pipe was cut perpendicular to the joint surface in order to observe the joint region and measure the flexural strength. Cross-section at the joint region was observed by optical microscopy. Flexural strength of the joined Si_3N_4 specimens was measured at room temperature using a four-point bending test on six specimens with dimension of 3 mm x 4 mm x 40 mm and with the joint region in the center of a bend bar.

RESULTS AND DISCUSSION

Figure 1 shows overview image of Si_3N_4 pipe joined at 1600°C for 1 h at 5 MPa. The Si_3N_4 long pipe of 2000 mm in length was successfully fabricated to join the two pipes of 1000 mm in length using the local heating.

Figure 2 shows cross-sectional image of the joint region in the Si_3N_4 long pipe joined at 1600°C by local heating. Neither break nor void was observed at the joint region by optical microscopy. The observation indicates that the melted insert filled the gap between the Si_3N_4 pipes. The thickness of the joining layer was about 10 m.

Table 1 shows flexural strength of Si_3N_4 specimens joined at 1600°C by local heating in comparison with that of the original Si_3N_4 bulk. The joined Si_3N_4 specimens had high joining strength of around 680 MPa in average value. The strength was similar to that of the original Si_3N_4 bulk. The joined specimens were fractured from joint region or bulk. The fracture from bulk indicates that strong joining layer and joint interface (between layer and bulk) would be formed at joining temperature of 1600°C by local heating.

Figure 1. Overview images of joined Si_3N_4 pipe of 2000 mm in length. The joined Si_3N_4 long pipe was obtained by joining two Si_3N_4 pipes of 1000 mm in length.

CONCLUSION

Local heating equipment was designed and developed to apply superplastic phenomenon of ceramics to the joining of tubular ceramics for the purpose of energy-saving joining. By using the local heating technique, a Si_3N_4 pipe of 2 meter long was produced by joining two Si_3N_4 pipes of 1 meter in length. The joined Si_3N_4 long pipe was successfully obtained without break and void at the joint region. Flexural strength of the joined Si_3N_4 pipe was similar to that of original Si_3N_4 bulk, indicating high joining strength of the joined Si_3N_4 long pipe by using local heating technique.

Figure 2. Cross-sectional image of the joint region in the joined Si_3N_4 long pipe.

Table 1. Flexural strength of joined Si_3N_4 specimen and original Si_3N_4 bulk.

Sample	Flexural strength (MPa)
Joined Si_3N_4 specimen	677 ± 61 (Max.: 727, Min.: 564)
Original Si_3N_4 bulk	737 ± 46

ACKNOWLEDGMENT

This work was supported by the Project for "Innovative Development of Ceramics Manufacturing Technologies for Energy Saving" from Ministry of Economy, Trade and Industry (METI), Japan and New Energy and Industrial Technology Development Organization (NEDO), Japan.

REFERENCES
1) M.A. Sainz, P. Miranzo and M.I. Osendi, Silicon Nitride Joining Using Silica and Yttria Ceramic Interlayers, *J. Am. Ceram. Soc.*, **85**, 941-46 (2002).
2) M. Gopal, M. Sixta, L.D. Jonghe and G. Thomas, Seamless Joining of Silicon Nitride Ceramics, *J. Am. Ceram. Soc.*, **84**, 708-12 (2001).
3) R.J. Xie, M. Mitomo, L.P. Huang and X.R. Fu, Joining of Silicon Nitride Ceramics for High-Temperature Applications, *J. Mater. Res.*, **15**, 136-41 (2000).
4) R.J. Xie, L.P. Huang, Y. Chen and X.R. Fu, Evaluation of Si_3N_4 Joints: Bond Strength and Microstructure, *J. Mater. Sci.*, **34**, 1783-90 (1999).

5) M. Nakamura and I. Shigematsu, Diffusion Joining of Si_3N_4 Ceramics by Hot Pressing under High Nitrogen Gas Pressure, *J. Mater. Sci. Lett.*, **16**, 1030-32 (1997).

6) H.L. Lee, S.W. Nam, B.S. Hahn, B.H. Park and D. Han, Joining of Silicon Carbide Using $MgO-Al_2O_3-SiO_2$ Filler, *J. Mater. Sci.*, **33**, 5007-14 (1998).

7) J.Q. Li and P. Xiao, Joining Alumina Using an Alumina/Metal Composite, *J. Eur. Ceram. Soc.*, **22**, 1225-33 (2002).

8) N. Kondo, H. Hyuga, H. Kita and K. Hirao, Joining of Silicon Nitride by Microwave Local Heating, *J. Ceram. Soc. Jpn.*, **118**, 959-62 (2010).

JOINING OF SILICON NITRIDE WITH PYREX® GLASS BY MICROWAVE LOCAL HEATING

Naoki KONDO, Hideki HYUGA, Mikinori HOTTA, Hideki KITA and Kiyoshi HIRAO
National Institute of Advanced Industrial Science and Technology (AIST)
Shimo-shidami 2266-98, Moriyama-ku, Nagoya 463-8560, Japan

ABSTRACT

Silicon nitride was joined using Pyrex® glass as an insert material. The glass powder was placed between the silicon nitride plates to make a joint. The glass joint was surrounded by silicon carbide granules. These were settled in a microwave furnace. Silicon carbide granules acted as susceptor and heated up by absorbing microwave. The glass powder was melted, and glass joint was formed between the silicon nitride plates. Strength of the joined silicon nitride was 110 MPa, therefore, good joining was achieved by microwave local heating.

INTRODUCTION

Joining technology is critically important for fabricating ceramic components. Silicon nitrides are one of the important ceramics for structural use, therefore, a large number of joining techniques have been developed for silicon nitride.[1] Joining techniques must be used depending on purposes. Glass insert is a candidate for the joint of silicon nitride, where moderate strength is required.

Microwave heating is an interesting heating technique, which can heat a material locally and rapidly. This technique was successfully applied to join silicon nitride with sialon glass joint.[2] For this joining, silicon carbide (SiC) was used as susceptor. SiC was located around the joint. It absorbed microwave and heated up, and local heating was achieved.

In this paper, we intended to join silicon nitride by Pyrex® glass insert. Advantages of the Pyrex® glass are 1) coefficient of thermal expansion is similar to that of silicon nitride,[3] and 2) softening temperature is relatively low. Microwave local heating was applied to melt the glass for joining. The microstructure of the joint was investigated and strength of the joined silicon nitride was examined.

EXPERIMENTAL

Silicon nitride used for this study was a commercially available one; whose composition was Si_3N_4 - 5wt.%Y_2O_3 - 5wt.%Al_2O_3. Coefficient of thermal expansion was 3.5×10^{-6} / °C. Plates of 5 x 20 x 25 mm were cut from a sintered block. Planes (5 x 20 mm) of the plates provided for joining were machined using a #400 whetstone.

Pyrex® glass used as an insert material was the Corning Pyrex® 7740. Nominal composition of the glass was 80.6 wt.% SiO_2 - 13.0 wt.% B_2O_3 - 4.0 wt.% Na_2O - 2.3 wt.% Al_2O_3 - 0.1 wt.% traces. Softening temperature was 821°C and coefficient of thermal expansion was 3.3×10^{-6} / °C.

Silicon carbide granules were used as susceptor, and alumina fiberboard was used as insulator. Silicon nitride plates, Pyrex® glass powder and alumina fiberboard are poor microwave absorbers at low temperatures. On the contrary, silicon carbide granules are microwave absorbers. Fig. 1 shows a schematic drawing of the configuration. Pyrex® glass powder was placed between two silicon nitride plates to make a joint. The joint was surrounded by silicon carbide granules. Height and thickness of the granules region were 20mm and 10mm, respectively. Therefore other sides of the silicon nitride plates were outside of the granules. As only silicon carbide granules were heated up by microwave radiation, local heating around the joint can be actualized. Above configuration was placed inside of the alumina fiberboard.

Microwave furnace used here was a commercially available microwave oven for kitchen use. One important purpose of this work was investigating the possibility of robust joining using Pyrex®

glass powder insert and microwave local heating. The authors could use better microwave furnace,[2] however, we dare to use the cheap microwave oven. Frequency and maximum output power of the oven were 2.45 GHz and 800 W. Microwave radiation was done in air condition, since oxidized surface was reported to be important for joining.[3] Maximum output power was applied to heat up. Only 5 min was needed to reach the maximum temperature. Temperature measurement by a pyrometer showed the maximum temperature reached more than 1300 °C. After holding at the maximum temperature for 1 min, output power was cut off. The joined silicon nitride was cooled below 500 °C in 10 min. No mechanical pressure was applied throughout the heating, holding and cooling procedures.

Specimens for microstructure observations and strength measurement were cut from the joined plate. Microstructure observations were performed by optical microscopy (OM) and scanning electron microscopy (SEM). Crystalline phases were examined by X-ray diffraction analysis. Specimens of 3 × 4 × 40 mm were prepared for strength measurement. These specimens have a joint at the center of a bend bar. Tensile side of the specimen was polished by diamond pastes to remove flaws introduced by machining. Four-point bending strength was measured in accordance with JIS R1601 using outer and inner spans of 30 and 10 mm, and a displacement rate of 0.5 mm min^{-1}.

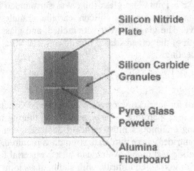

Silicon Nitride
Plate

Silicon Carbide
Granules

Pyrex Glass
Powder

Alumina
Fiberboard

Fig. 1
A schematic view from the side, showing the configuration of silicon nitride plates, pyrex glass powder, silicon carbide granules and alumina fiberboard.

RESULTS AND DISCUSSION

After the microwave heating, Pyrex® glass power melted to be a bulk glass. The glass flowed out from the joint. Fig. 2 shows the glass spread on the silicon nitride surface. As joining was done under air condition and more than 1300 °C, surface of the silicon nitride was oxidized and covered by glass. Needle-like crystalline phase was observed in the glass, which was confirmed to be cristobalite by X-ray diffraction analysis. Softened Pyrex® glass flowed out from the joint, and spread on the oxidized silicon nitride surface. Separation of the Pyrex® glass from the silicon nitride surface could not be observed. It was difficult to measure the wet contact angle, however, the Pyrex® glass certainly wetted the silicon nitride surface and made a bond. Many cracks were found in the flowed out Pyrex® glass. Cooling rate was too fast; this led to cracking in the Pyrex® glass.

Specimen for microstructure observation was cut from the joined plate. Thickness of the glass joint between the silicon nitride plates was about 30μm. Fig. 3 shows a SEM micrograph of the joined interface between the silicon nitride and the Pyrex® glass, which was taken from the polished surface. Fig. 4 shows an optical micrograph taken from the bend specimen after strength measurement.

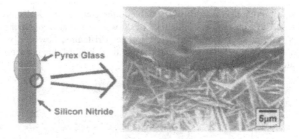

Fig. 2   Edge of the Pyrex® glass on silicon nitride surface.
Schematic drawing shows the point where SEM observation was carried out.
Upper and lower sides of the SEM micrograph are Pyrex® glass
and silicon nitride surface, respectively.
Softened Pyrex® glass flowed out from the joint, and spread
on the silicon nitride surface.

Pyrex® glass successfully filled the gap between the two silicon nitride plates. Neither pores (bubbles) nor cracks were observed in the Pyrex® glass joint. The crack in the Pyrex® glass flowed out on the surface did not extend into the glass joint. A groove was observed at the interface in Fig. 3. This groove was most likely caused by polishing. Diamond paste preferentially attacked the interface, resulted in formation of the groove. This was considered from the following two reasons. First, no infiltration of ink into the interface was observed by ink check. Second, crack did not run through the interface after bend test, but run through the center of the glass joint. If the groove was a crack or separation, the crack must run through the interface.

No crystalline phase except for silicon nitride was found around the joint interface as shown in Fig. 3. This was also confirmed by X-ray diffraction analysis, which was performed on the fractured surface of the bend specimen.

Fig. 3
Joined interface between silicon nitride and pyrex glass.
Micrograph was taken from the polished surface.

Kalemtas et. al., examined the joined interface between pre-oxidized SiAlON and Pyrex® glass.[3] Cristobalite phase formation at the interface in the Pyrex® glass side was reported, when cooling rate after joining was slow (~ 40°C /min). Concentration of Y at the interface due to diffusion from SiAlON was also reported, when time and temperature for pre-oxidation of SiAlON were long (> 2h) and high (> 1000 °C). They concluded that minimum pre-oxidation of SiAlON and rapid cooling were important to obtain good joint interface, and they demonstrated it.

The joining condition of this work was suitable to the condition by Kalemtas et. al. to obtain good joint. Neither cristobalite nor concentration of Y was observed around the joint interface. Concentration of Y was prevented by short heating time. Cristbalite, however, was found on the oxidized surface of silicon nitride. Not only rapid cooling time, but also the elements in Pyrex® glass suppress the formation of cristbalite.

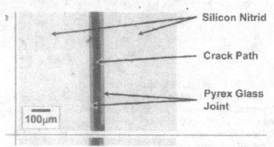

Fig. 4   Crack path of the joined specimen after strength measurement.
The crack mainly run center of the glass joint.

Strength of the joined specimen was 110 MPa. This was low compared to the previous works joined by SiAlON glass.[2, 4] Strength of the Pyrex® glass itself was lower compared to the SiAlON glass, resulted in the lower strength of joined specimen. Still the joining by Pyrex® glass using microwave local heating in air condition is an important technique, since the technique is much easier to make a joint compared to the technique using SiAlON glass. Moderate strength of this work is enough to make airtight joint, for instance.

CONCLUSIONS
Joining of silicon nitride plates by Pyrex® glass was intended. Microwave local heating was applied to melt the glass. Melted glass filled the gap between the silicon nitride, and glass joint was successfully formed. Neither pores (bubbles) nor cracks were observed in the Pyrex® glass joint. No crystalline phase except for silicon nitride was found around the joint interface. Strength of the joined silicon nitride was 110 MPa, therefore, good joining was achieved by microwave local heating.

REFERENCES
[1] Loehman RE, "Recent Progress in Ceramic Joining," *Key Eng. Mater.*, **161-163**, 657-661, (1999).
[2] Kondo N, Hyuga H, Kita H, Hirao K, "Joining of silicon nitride by microwave local heating", *J. Ceram. Soc. Japan.*, **118**, 959-962, (2010).
[3] Kalemtas A, Kara A, Kara F, Mandal H, Aktug B, "Joining of SiAlON ceramics to Pyrex® glass," *Silicates Industries*, **69**, 219-224, (2004).
[4] Johnson M and Rowcliffe DJ, "Mechanical properties of joined silicon nitride," *J. Am. Ceram. Soc.*, **68**, 468 472, (1985).

ACKNOWLEDGEMENT
This research was supported by METI and NEDO, Japan, as part of the Project "Development of innovative ceramics manufacturing technologies for energy saving."

MECHANICAL PROPERTIES OF CHEMICAL BONDED PHOSPHATE CERAMICS WITH FLY ASH AS FILLER

H. A. Colorado[1,2], C. Daniel[3], C. Hiel[4,5], H. T. Hahn [1,3], J. M. Yang[1]

[1]Materials Science and Engineering Department, University of California, Los Angeles, CA 90095, USA; E-mail: hcoloradolopera@ucla.edu
[2]Universidad de Antioquia, Mechanical Engineering. Medellin-Colombia
[3]Mechanical and Aerospace Engineering Department, University of California, Los Angeles
[4]Composite Support and Solutions Inc. San Pedro, California
[5]Mechanics of Materials and Constructions. University of Brussels (VUB)

ABSTRACT

This paper is concerned with the use of fly ash filler in chemically bonded phosphate ceramic (CBPC) composites, with the goals of reducing production costs and developing an environmentally benign structural material. Traditional ceramics are usually associated with high temperature processing which increases global warming. Fortunately, CBPC manufacturing can be performed at room temperature and the material itself is biocompatible.

The CBPC was produced by mixing a phosphoric acid formulation and a controlled Wollastonite powder (both from Composites Support and Solutions). Fly ash was used as a filler material. Compressive strength, microstructure and different manufacturing process parameters were evaluated. Microstructure was identified by using optical and scanning electron microscopy and X-ray diffraction. The microstructure characterization shows that CBPC is a composite itself with several crystalline (Wollastonite and brushite) and amorphous phases.

INTRODUCTION

Chemically Bonded Ceramics (CBCs) are inorganic solids synthesized by chemical reactions at low temperatures without the use of thermally activated solid-state diffusion (typically less than 300°C). This method avoids high temperature processing (by thermal diffusion or melting) which is the nomal in traditional ceramics processing. The chemical bonding in CBC's allows them to be inexpensive in high volume production. Because of this, CBCs have been used for multiple applications. They include: dental materials [1], nuclear waste solidification and encapsulation [2], electronic materials [3], composites with fillers and reinforcements ([4] and [5]). The fabrication of conventional cements and ceramics is energy intensive as it involves high temperature processes and emission of greenhouse gases, which adversely affect the environment.

In this research, a CBC formed by Wollastonite powder ($CaSiO_3$) and phosphoric acid (H_3PO_4) was used, which, when mixed in a ratio of 100/120 reacts into a Chemically Bonded Phosphate Ceramic (CBPC). In CBPCs, when the aqueous phosphoric acid formulation and the Wollastonite powder mixture are stirred, the sparsely alkaline oxides dissolve and an acid base reaction is initiated. The result is a slurry that hardens into a ceramic product [6]. Wollastonite is a natural calcium meta-silicate which is mostly used as a filler in resins and plastics, ceramics, metallurgy, biomaterials and other industrial applications [7].

The mixing of Wollastonite with phosphoric acid produces calcium phosphates (brushite ($CaHPO_4 \cdot 2H_2O$), monetite ($CaHPO_4$) and calcium dihydrogenphosphate monohydrate ($Ca(H_2PO_4)2 \cdot H_2O$)) and silica for molar ratios (P/Ca) between 1 and 1.66 [8]. For comparison, sintered ceramics of Wollastonite powders (consolidated by applying pressure and sintering as opposed to the CBPC by chemical reactions) have been evaluated. CBPC matrix composites with Fly ash (class C and F) as filler have been fabricated before [5] for non-Wollastonite based CBPCs: MgO mixed with the acid KH_2PO_4 produced $MgKPO_4 \cdot 6H2O$. For class F the compressive strength increases from 23.4 Mpa for

CBPC to 52.5 MPa for CBPC with 50% Fly ash. However, no further details are reported about the standard deviation or variability of data.

The sections below will present the compression strength and curing properties of wollastonite-based CBPCs with Fly ash class F as filler. Compression tests were conducted for both CBPCs and sintered samples to determine their respective mechanical strengths.

EXPERIMENTAL
CBPC manufacturing

The manufacturing of CBPC samples was conducted by mixing an aqueous phosphoric acid formulation (from Pastone USA) and natural Wollastonite with Fly ash powders in a 1.2 to 1.0 ratio of liquid to powder. The pH of the CBPC after curing was 7.0. The composition of Wollastonite (from Minera Nyco) and Fly ash class F (from Diversified minerals Inc) are presented in Table 1 and 2.

Table 1 Chemical composition of Wollastonite powder.

Composition	CaO	SiO₂	Fe₂O₃	Al₂O₃	MnO	MgO	TiO₂	K₂O
Percentage	46.25	52.00	0.25	0.40	0.025	0.50	0.025	0.15

Table 2 Chemical composition range of Fly ash class F

Composition	CaO	SiO₂	Fe₂O₃	Al₂O₃
Percentage	5-22	59-63	2-5	11-15

Curing and compression samples were fabricated. The mixing process of Wollastonite powder and phosphoric acid formulation was done in a Planetary Centrifugal Mixer (Thinky Mixer® AR-250, TM). The mixture of Wollastonite powder and the phosphoric acid formulation were maintained at room temperature to cure. For compression samples, the mixture of Wollastonite powder and the phosphoric acid formulation were maintained at 3 °C in a closed container (to prevent water absorption) for 1 hour in order to increase the pot life of the resin.

Curing tests

For all cure samples, 24g of the acid solution and 20g of powder (Wollastonite + fly ash) were mixed at room temperature in the TM for 15 sec in a container with a hole to allow resulting gases to escape. Three factors were considered in determining this short mixing time:

   some Wollastonite powders are very reactive due to their grain sizes;

   the small amount of total precursor (44g) requires short mixing time; and

   at room temperature it is easier to see the effect of using different powders , additives and processing methods on the setting time.

Curing curves were then obtained by measuring the temperature change with time for different filler content. Both mixing and curing were conducted in TM containers. The mixing was performed in 125ml Polypropylene jars while the curing was performed in 24ml Polypropylene jars. A thermocouple was set at the bottom of the container, as shown in Figure 1a. The same containers (used for mixing and for curing) were used following the same parameters and experimental conditions for all samples.

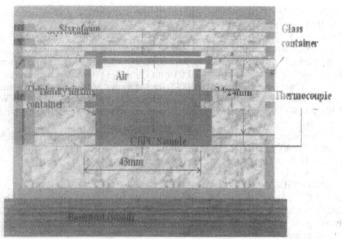

Figure 1 a) Experimental set up for the curing experiments. All tests were conducted at room temperature in a closed (to humidity) plastic container, with a thermocouple inserted into a hole and then sealed with tape.

Compression tests

For all compression samples, 120g of the acid solution and 100g of Wollastonite, both at 3 °C, were mixed in the TM in a container with a hole in the cap to allow for the release of upcoming gases. The mixing time was 3 min for CPBC; 2 and 3 min for CBPC with 1.0wt% of Fly ash; 2 min for CBPC with 10wt% of Fly ash; 25 sec for CBPC with 50wt% of Fly ash; and 1 min for CBPC with 50wt% of Fly ash. The maximum mixing time (related to the setting time) is inversely related to the Fly ash content, so less mixing time is required as Fly ash content is increased.

A Teflon® fluoropolymer mold with mold release (Synlube 1000 silicone-based release agent applied before the mixture discharge) was used to minimize the adhesion of the CBPC to the mold. Next, the mold with the CBPC was covered with plastic foil to prevent exposure to humidity and decrease shrinkage effects. Samples were released after 48 hours and then dried at room temperature in open air for at least 3 days. Samples were then mechanically polished with parallel and smooth faces (top and bottom) for the compression test. Since the CBPC has both unbonded and bonded water, samples were dried slowly in the furnace in order to prevent residual stresses first at 50°C for 1 day, followed by 105°C for an additional day.

Compression tests were conducted in an Instron® machine 3382, over cylindrical CBPC samples (9mm in diameter by 20mm in length) for M200, M200 with 1, 10, 20 and 50% of fly ash. Also, the effect of the mixing time on the compression strength was studied for the CBPC with 1% of fly ash. A set of 20 samples were tested for each powder. The crosshead speed was 1mm/min.

Other Characterization:

To see the microstructure, sample sections were ground using silicon carbide papers of 500, 1000, 2400 and 4000 grit progressively. Then they were polished with alumina powders of 1, 0.3 and 0.05µm grain size progressively. After polishing, samples were dried in a furnace at 70°C for 4 hours Next, samples were mounted on an aluminum stub and sputtered in a Hummer 6.2 system (15mA AC for 30 sec) creating a 1nm thick film of Au. The SEM used was a JEOL JSM 6700R in a high vacuum mode. Elemental distribution x-ray maps were collected on the SEM equipped with an energy-dispersive analyzer (SEM-EDS). The images were collected on the polished and gold-coated samples, with a counting time of 51.2 ms/pixel.

X-Ray Diffraction (XRD) experiments were conducted usin X'Pert PRO equipment (Cu Kα radiation, λ=1.5406 Å), at 45KV and scanning between 10° and 80°. M200 and M200 with 1, 10, 20 and 50% of fly ash samples were ground in an alumina mortar and XRD tests were done at room temperature.

Finally, density tests were conducted over CBPCs with Fly ash as filler. All samples were tested after a drying process (50°C for 1 day, followed by 100°C for 1 day) in a Metter Toledo™ balance, by means of the buoyancy method. Six samples for each composition were tested. The Dry Weight (Wd), Submerged Weight (Ws), and Saturated Weight (Wss) were measured. The following parameters were calculated:

Bulk volume: $Vb = Wss - Ws$; Apparent volume: $Vapp = Wd - Ws$; Open-pore volume: $Vop = Wss - Wd$; % porosity = $(Vop/Vb) \times 100$ %; Bulk Density: $Db = Wd/(Wss - Ws)$; and Apparent Density: $Da = Wd/(Wd - Ws)$. In these calculations, density of water was taken to be 1.0 g/cm^3.

ANALYSIS AND RESULTS

Typical acicular Wollastonite and Fly ash microparticles are presented in Figure 2a and b respectively. A magnification of a typical Fly ash particle showing a rough surface is shown in Figure 2c. Figure 2d shows a Fly ash particle in a CBPC matrix (with 1.0 wt% of Fly ash). Microcracks appeared at the interface particle-matrix.

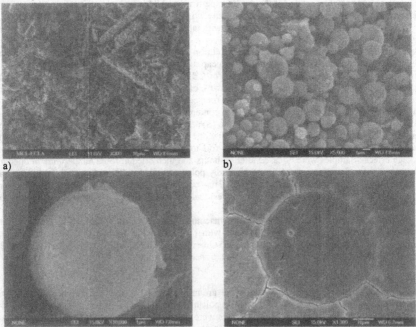

a) b) c) d)
Figure 2 SEM cross section view images of a) M200 Wollastonite powder, b) Fly ash class F, c) Magnification of a Fly ash particle from b), d) detail of CBPC showing a Fly ash particle in a CBPC matrix with 1.0 wt % of Fly ash as filler.

Figure 3a shows X-ray maps for a CBPC with 50% of Fly ash as filler. In general, the distribution of P, Ca and C is homogeneous everywhere except in the Ash particles, where Ca dominates.

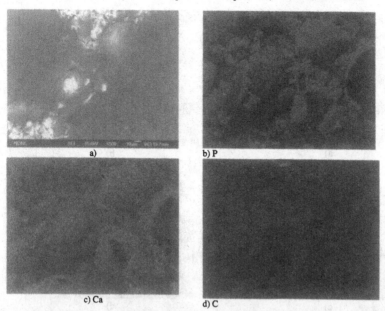

Figure 3 X-Ray maps for a CBPC with 50 wt % of Fly ash class F as filler, a) Mophology, b) P content, c) Ca content, d) C content.

Figure 4a shows a CBPC with 1.0wt % of Fly ash as filler. The particle of about 15µm in diameter is stopping cracks, acting as reinforcement. Some of these particles have big voids. A magnification of what is presented in Figure 4b. Half of the particle-CBPC interface has a layer of about 1µm produced during the reaction between Fly ash and phosphoric acid. This product seems to be mainly calcium phosphates and maybe amorphous calcium silicate hydrate (C–S–H), produced from the reaction of the amorphous silica from the Fly ash with sub-products like $Ca(OH)_2$. Figure 4c on the other hand shows a CBPC with 50wt % of Fly ash as filler. Due to the high loading of filler, the CBPC matrix does not appear as smooth as the 1.0wt % sample, and many Fly ash particles appear with holes approximately the size of the particle diameter, with some new materials inside. This can indicate a very strong reaction between some ash particles and phosphoric acid enables to penetrate the hollow particle when the wall is thin. Figure 4d supports this idea, showing a magnification of the interface where particles are broken in some points from which new phases like phosphates grow (right part of the image corresponds to the interior of the sphere). X-ray maps (from Figure 3 and other experiments not presented) seem to indicate that the interior of the microsphere in the 50wt % of Fly ash loading is constituted mainly by calcium (either calcium hydroxides or quick lime). More research is being conducted to completely identify these products. Figure 5 shows curing curves of CBPCs fabricated with M200 Wollastonite powders at room temperature and with Fly ash class F as filler in different concentrations. Figure 5a shows that when the Fly ash content is increased, the curing curve is moved to the left, which means, the curing time is decreased.

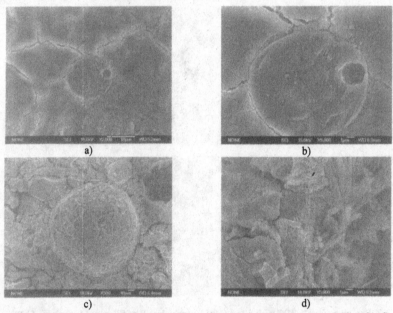

Figure 4 CBPC with Fly ash as filler; a) 1wt% and b) magnification of a); c) 50 wt %, d) magnification of the interface Fly ash-CBPC for the biggest particle from image (left-center side).

Figure 5b shows the setting time for the CBPC with different Fly ash content, calculated from the inflexion points of curves in Figure 5a.

It has been established with other cements like Portland that Fly ash particles behave more or less as an inert material and serve as nuclei for precipitation of calcium hydroxides and amorphous calcium silicate hydrate (C–S–H). A similar but also stronger effect is possible with CBPCs, evidenced by the decrease in the setting times when Fly ash content increases.

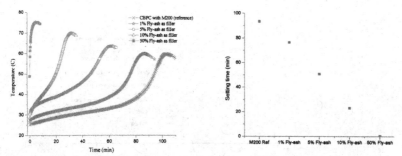

Figure 5 Curing for CBPC fabricated at room temperature with M200 wollastonite powder; a) curing curves for CBPC with Fly ash as filler, b) Setting time estimated from a).

Figure 6 shows the XRD for Fly ash, CBPC fabricated with M200 Wollastonite powder and CBPCs with Fly ash as filler. Figure 6a shows that in general Fly ash is difficult to identify in the CBPC, with only slight differences in intensity, even with 50wt% as filler. This can be explained from composition results presented in Table 1 and 2, in which the first four constituents of Wollastonite (CaO, SiO_2, Fe_2O_3, Al_2O_3), which make up 98.9% of the material, are also the major contents present in the Fly ash. This means that almost the same products can be obtained when these powders react with the phosphoric acid. Also, it is observed that mixing time does not have significant effects in the composition of the composite.

Figure 6b shows only three selected spectra from Figure 6a, Fly ash, CBPC and CBPC with 50wt% of Fly ash. A peak at 50° is present in the Fly ash and in the CBPC with 50wt% of filler and is almost the only difference between the CBPC without filler and the CBPC with 50% of filler. Other diffraction peaks either are masked by other peaks or do not appear.

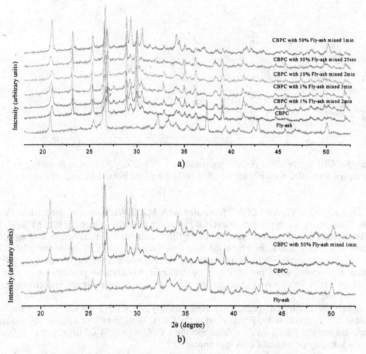

Figure 6 XRD for Fly ash, CBPC and CBPCs with Fly ash as filler.

Finally, compression results are summarized on Table 3. The Fly ash content has a significant effect on CPBCs. The mean of compressive strength decreases as Fly ash content increases. Also, as mixing time increases, the mean of compressive strength increases. This can be explained as a result of better mixing, which can produce more homogeneous material, as well as a material with less voids.

Table 3 Summary of compression test results for the CBPCs with Fly ash as filler.

Sample	Minimum Strength (MPa)	Maximum Strength (MPa)	Mean value (MPa)	Standard deviation (MPa)
CBPC and 3 min mixing	78.0	116.7	102.3	10.4
CBPC with 1% Fly ash and 2min mixing	50	94	72.1	13.5
CBPC with 1% Fly ash and 3min mixing	69	105	85.1	9.4
CBPC with 10% Fly ash and 2min mixing	37	98	60.2	21.4
CBPC with 50% Fly ash and 25sec mixing	2.6	9	5.1	2.2
CBPC with 50% Fly ash and 1min mixing	8.4	17.2	12.3	2.3

Figure 8 shows the error bars for the data presented in Table 3. It can be seen how the mean compressive strength is increased with mixing time.

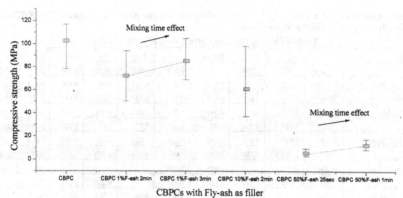

Figure 7 Compressive strength for CBPCs with Fly ash as filler.

Figure 8 shows the Weibull distributions for CBPCs (fabricated with Wollastonite powder M200) with Fly ash class F as filler. The biggest values were obtained for the CBPC with no filler.

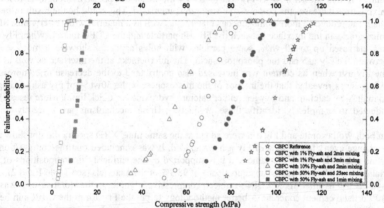

Figure 8 Weibull distributions for CBPCs (fabricated with Wollastonite powder M200) with Fly ash as filler.

Table 4 shows the results of density tests over CBPCs with Fly ash as filler. The bulk density and the percentage of porosity slightly decreased when mixing time was increased from 2 to 3 min for the CBPC with 1wt% Fly ash. This can be associated with a larger amount of amorphous phases, which can hold significant amounts of micro- and nano-pores. This result is interesting since it was shown in Figure 7 that compressive strength increased when mixing time increased, which opens up applications involving light ceramic materials. Also, for the CBPC with 10wt% Fly ash mixed for 2min the bulk density as well as the porosity increased when they are compared with 1wt% Fly ash mixed for 2min. When Fly ash content was increased up to 50wt%, however, the bulk density had the lowest value when mixing time was 25 sec. It reached the highest density value when mixing time was 1 min. This suggests a more complex reaction, involving mainly amorphous phases not detected by XRD. More research is being conducted on these results.

Table 4 Density tests for CBPCs with Fly ash as filler

Sample		Wd (g)	Ws (g)	Wss (g)	Vb	Vapp	Vop	Db	Da	% por
CBPC with 1wt% Fly ash mixed 2min	Mean	1.807	0.966	2.058	1.092	0.840	0.252	1.654	2.150	23.083
	St. Dev.	0.058	0.024	0.049	0.025	0.034	0.009	0.016	0.018	1.343
CBPC with 1wt% Fly ash mixed 3min	Mean	1.588	0.855	1.841	0.986	0.733	0.253	1.611	2.167	25.660
	St. Dev.	0.031	0.017	0.041	0.024	0.014	0.010	0.008	0.005	0.376
CBPC with 10wt% Fly ash mixed 2min	Mean	1.703	0.931	1.980	1.049	0.772	0.277	1.623	2.206	26.415
	St. Dev.	0.042	0.019	0.043	0.024	0.023	0.002	0.007	0.014	0.471
CBPC with 50wt% Fly ash mixed 25sec	Mean	1.467	0.758	1.690	0.932	0.709	0.223	1.576	2.073	23.957
	St. Dev.	0.075	0.016	0.075	0.059	0.059	0.001	0.020	0.067	1.499
CBPC with 50wt% Fly ash mixed 1min	Mean	1.586	0.872	1.802	0.930	0.714	0.216	1.706	2.222	23.234
	St. Dev.	0	0.001	0.001	0.001	0.001	0.001	0.002	0.002	0.084

SUMMARY
The effect of Fly ash class F as filler in the compressive strength of Wollastonite based-CBPCs has been presented. It was shown how as Fly ash increases the mean compressive strength is reduced. Also, as mixing time increases the compressive strength is increased. This could be the result of better mixing, which can produce a more homogeneous material, as well as a material with fewer voids. SEM revealed microcracks in the interface between the Fly ash particle and the CBPC matrix. When Fly ash content was increased up to 50 wt%, some particles with holes appeared showing a more severe reaction between the Fly ash and the phosphoric acid. The microcracks at the interface as well as the holes in the Fly ash when its content was increased also contributed to the decrease in compressive strength. X-ray maps revealed that the interior of the microsphere in the 50wt % of Fly ash loading is constituted mainly by calcium and oxygen (either calcium hydroxides or quick lime). More research is being conducted to completely identify these products. These mechanisms are currently under research.

Even when both Wollastonite and Fly ash were added at the same time, XRD spectra did not show any significant change even after 50wt% of Fly ash was added. It was concluded that the composition of the crystalline phases is almost the same and it is supported by the similarity in compositions of the raw materials, for which there can be an equivalence of 98.9% of the materials (see Table 1 and 2).

Even though Fly ash as filler in the CBPCs reduced compressive strength, in the worst cases it was on the order of Portland cement concrete. A better methodology to put the Fly ash in the CBPCs in being conducted, not only to increase the setting time but also to decrease the adverse severe reaction of an acid-filler. Results will be presented in a future paper.

Weibull distributions for different CBPCs with Fly ash as filler showed the variability of compressive strength over a set of 20 samples for each composition. This data shows the lower limits which are important as design criteria. Currently different manufacturing processing are being worked on in order to increase these minimum values as well as the variability.

Finally, density results showed a complex microstructure-processing relationship. At 1wt% Fly ash, when mixing time was increased, the bulk density decreased. On the contrary, at 50wt% Fly ash, when mixing time was increased, the bulk density increased. More research is being conducted on this topic.

ACKNOWLEDGEMENTS
The authors wish to thank to the NIST-ATP Program through a grant to Composites and Solutions Inc. (Program Monitor Dr. Felix H. Wu) and to Colciencias from Colombia for the grant to Henry A. Colorado.

REFERENCES
1. L. C. Chow and E. D. Eanes. Octacalcium phosphate. Monographs in oral science, vol 18. Karger, Switzerland, 2001.
2. D. Singh, S. Y. Jeong, K. Dwyer and T. Abesadze. Ceramicrete: a novel ceramic packaging system for spent-fuel transport and storage. Argonne National Laboratory.Proceedings of Waste Management 2K Conference, Tucson, AZ, 2000.
3. J. F. Young and S. Dimitry. Electrical properties of chemical bonded ceramic insulators. J. Am. Ceram. Soc., 73, 9, 2775-78 (1990).
4. T. L. Laufenberg, M. Aro, A. Wagh, J. E. Winandy, P. Donahue, S. Weitner and J. Aue. Phosphate-bonded ceramic-wood composites. Ninth International Conference on Inorganic bonded composite materials (2004).
5. A. S. Wagh. Chemical bonded phosphate ceramics. *Elsevier* Argonne National Laboratory, USA. 283 (2004).
6. S. Y. Jeong and A. S. Wagh. Chemical bonding phosphate ceramics: cementing the gap between ceramics, cements, and polymers. Argonne National Laboratory report, June 2002.
7. T.K. Kundu, K. Hanumantha Rao, S.C. Parker. Atomistic simulation of the surface structure of Wollastonite and adsorption phenomena relevant to flotation. Int. J. Miner. Process. 72, 111 –127 (2003).
8. Mosselmans G, Monique Biesemans, Willem R, Wastiels J, Leermakers M, Rahier H, Brughmans S, and Van Mele B 2007 Journal of Thermal Analysis and Calorimetry Vol. 88 3 723.
9. Effect of curing conditions and ionic additives on properties of fly ash–lime compacts, Saikat Maitra, Farooq Ahmad, Ananta K Das, Santanu Das and Binay K Dutta. Bull. Mater. Sci., Vol. 33, No. 2, April 2010, pp. 185–190.
10. Colorado H. A, Hahn H. T. and Hiel C. Pultruded glass fiber- and pultruded carbon fiber-reinforced chemically bonded phosphate ceramics. To appear at Journal of Composite Materials. Manuscript ID JCM-10-0398, (2010).

CERAMICS MICRO PROCESSING OF PHOTONIC CRYSTALS: GEOMETRICAL PATTERNING OF TIANIA DISPERSED POLYMER FOR TERAHERTZ WAVE CONTROL

Soshu Kirihara, Satoko Tasaki
Smart Processing Research Center
Joining and Welding Research Institute
Osaka University
11-1 Mihogaoka Ibaraki, Osaka 567-0047, Japan

Toshiki Niki, Noritoshi Ohta
Division of Sustainable Energy and Environmental Engineering
Graduate School of Engineering
Osaka University
2-1 Yamadaoka Suita, Osaka 565-0871, Japan

ABSTRACT

Micro patterns with periodic arrangements of dielectric polygon tablets were designed and fabricated successfully to control microwave energy concentrations and propagations in terahertz frequency ranges by using stereolithography. Photo sensitive resins including titania particles were spread on a glass substrate, and ultra violet images were exposed by digital micro-mirror device. Electromagnetic wave properties were measured by using a terahertz spectroscopic device. In transmission spectra of the dielectric patterns, widely forbidden bands and sharply localized mode were observed. Multiple reflection modes in periodic patterns were visualized by using transmission lime modeling. In the terahertz wave frequency range, the microwave can harmonize with vibration modes of saccharide or protein molecules. These micro patterns will be applied to censors or reactors to detect and create the biochemical materials through the terahertz wave excitations.

INTRODUCTION

Photonic crystals with periodic arrangements of dielectric materials can exhibit forbidden gaps in electromagnetic wave frequency ranges through Bragg diffraction [1-4]. Opaque regions especially called photonic band gap are formed in transmission spectra at wavelength ranges corresponding to the lattice constant of the artificial crystals. By introduction of point or plane defects, localized modes with sharp transmission peaks can be observed in the band gaps [5-6]. Incident electromagnetic waves are resonated and confined in the defect regions through multiple reflections between the periodic structures. During the last decade, various radio and optical wave devices of novel antennas, filters and resonators

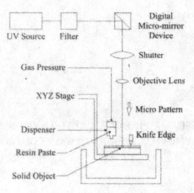

Figure 1 A schematically illustrated free forming system of a micro-stereolithography machine by using computer aided design and manufacturing (CAD/CAM) processes. (D-MEC Co. Ltd., Japan, SI-C 1000, http://www. d-mec.co.jp).

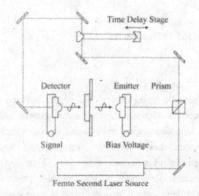

Figure 2 The schematically illustrated measuring system of a terahertz wave analyzer by using a time domain spectroscopic (TDS) detect method (Advanced Infrared Spectroscopy, Co. Ltd. Japan, J-Spec 2001, http://www. aispec.com).

hadbeen developed by utilizing the photonic crystals including the various structural defects with intentions to control the electromagnetic wave propagation, select the frequency, and localize the energy. In our previous investigations, the millimeter order photonic crystals were fabricated to control the microwaves with the gigahertz (GHz) frequencies by using laser scanning stereolithography of a computer aided design and manufacturing (CAD/CAM) system [7-9]. Recently, terahertz (THz) waves of unexplored microwaves with the wavelength from 30 μm to 3 mm and the frequencies form 100 GHz to 10 THz are expected to be applied for novel sensor to detect fine cracks in materials surfaces and small defects in electric circuits, and to analyze cancer cells in human skins and bacterium in foods. In our investigation group, an advance system of micro pattering stereolithography was newly developed to create the photonic crystals for the terahertz wave controls [10-14]. The micro photonic crystals composed of ceramic lattices with or without the structural defects were fabricated successfully [15-23]. In this investigation, the flat photonic crystals with the periodic arrangements of dielectric tablets composed of acrylic·resins including titania particles were fabricated to realize the effective wave diffractions, resonations and localizations in the terahertz frequencies. Filtering effects of the electromagnetic waves for a perpendicular direction to the dielectric patterns were observed through spectroscopic measurements. These micro geometric patterns of extremely thin devices with a high dielectric constant and permeability were designed to concentrate the terahertz wave energies effectively through theoretical simulations.

EXPERIMENTAL PROCEDURE

The micro dielectric pattern was designed as the periodic structure composed of square and hexagonal tablets of 240 and 120 μm in edge length at intervals of 45 and 60 μm, respectively. These squire and hexagonal tablets of $9 \times 9 = 81$ in numbers were arranged to form the extremely thin dielectric devices of 100 and 200 μm in pattern thickness, respectively. Real samples were fabricated by using the micro patterning stereolithography system. Designed graphic models were converted for stereolithography (STL) data files and sliced into series of two dimensional layers. These numerical data were transferred into the stereolithographic equipment (SIC-1000, D-MEC, Japan). Figure 1 shows a schematic illustration of the fabrication system. As the raw material, nanometer sized titania particles of 270 nm in average diameter were dispersed into a photo sensitive acrylic resin at 40 volume percent. The mixed slurry was squeezed on a working stage from a dispenser nozzle. This material paste was spread uniformly by a moving knife edge. Layer thickness was controlled to 10 μm. Ultra violet lay of 405 nm in wavelength was exposed on the resin surface according to the computer operation. Cross sectional layers of solid patterns were obtained by a light induced photo polymerization. High resolutions in these micro patterns had been achieved by using a digital micro mirror device (DMD). In this optical device, square aluminum mirrors of 14 μm in edge length were assembled with 1024×768 in numbers. Each micro mirror can be tilted independently, and cross

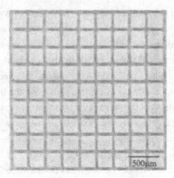

Figure 3   A dielectric micro pattern composed of titania particle dispersed acrylic resin fabricated by using the micro-stereolithography.

Figure 4   Microstructure in an acrylic resin bulk with titania particle dispersion observed by using a scanning electron microscope (SEM).

sectional patterns were dynamically exposed through objective lenses as bitmap images of 2 μm in space resolution. After stacking and joining these layers through photo solidifications, the periodical arrangements of the micro dielectric tablets were obtained. The titania particles dispersion in the acryl tablets were observed by using a scanning electron microscope (SEM). A terahertz wave attenuation of transmission amplitudes through these micro periodic patterns were measured by using a terahertz time domain spectrometer (TDS) apparatus (Pulse-IRS 1000, AISPEC, Japan). Figure 2 shows the schematic illustration of the measurement system. Femto second laser beams were irradiated into a micro emission antenna formed on a semiconductor substrate to generate the terahertz wave pulses. The terahertz waves were transmitted through the micro patterned samples perpendicularly. The terahertz wave diffraction and resonation behaviors in the micro patterns were calculated theoretically by using a transmission line modeling (TLM) simulator (Micro-stripes Ver. 7.5, Flomerics, UK) of a finite difference time domain (FDTD) method.

RESULTS AND DISCUSSION

The dielectric micro patterns with the periodic arrangement of the square acryl tablets with the titania particles dispersion was fabricated successfully by using the micro stereolithography system as shown in figure 3. Dimensional accuracies of the fabricated micro tablets and the air gaps were approximately 0.5 percent in length. These nanometer sized titania particles were verified to disperse uniformly in the acrylic resin matrix thorough the SEM observation as shown in figure 4. The dielectric constant of the composite material of the titania dispersed acrylic resin was measured as 40. Figure 5-(a) and (b) show transmission spectra measured and simulated by using the TDS and TLM methods, respectively. The measured result has good agreement with the calculated one. Opaque regions were formed in both spectra form 0.33 to 0.53 THz approximately. Maximum attenuation was measured as about -20 dB in transmission amplitude, and the minimum transmittance showed below 1 percent. The two dimensional photonic crystals with periodic arrangement with the lower dielectric contrasts were well known to open the band gaps limitedly for the parallel directions to the plane structures. However, the micro patterns with the periodically arranged square tablets above 30 in dielectric constant could be verified to exhibit the clear forbidden bands in the transmission spectra toward the perpendicular direction to the plane patterns through the theoretical simulations. The fabricated dielectric pattern is considered to totally reflect the terahertz wave at the wavelength comparable to the optical thickness as schematically illustrated in figure 6. Two different standing waves vibrating in the air and the dielectric regions form the higher and the lower edges of the band gap, respectively. The gap width can be controlled by varying geometric profile, filling ratio, and the dielectric constant of the tablets. As shown in figure 5, the localized mode of a transmission peak was observed at 0.47 THz in the band gap. Figure 7 shows a simulated distribution of electric field intensities in the micro pattern at the localized frequency. The white area indicates that the electric field

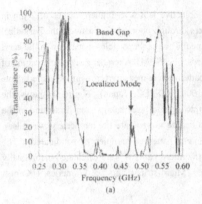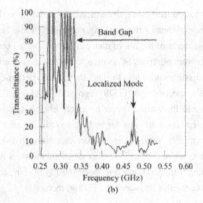

Figure 5 Transmission amplitudes of the terahertz wave through the dielectric micro pattern. The spectra (a) and (b) are measured and calculated properties by using the terahertz wave time domain spectroscopy (TDS) and a transmission line modeling (TLM) methods, respectively. In both transmission spectra, localized modes of transmission peaks are formed at specific frequencies in band gap regions.

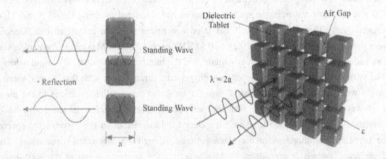

Figure 6 The schematic illustrations of formation mechanisms of the electromagnetic band gap through the dielectric micro pattern by Bragg diffraction. Incident direction of the electromagnetic wave is perpendicular to the dielectric arrangement with the plane structure.

intensity is high, whereas the black area indicates it is low. The incident terahertz wave is resonated and localized along the two dimensionally arranged dielectric tablets at the specific frequency. The amplified terahertz wave can transmit through the micro pattern. Therefore, the transmission peak of the localized mode should be formed clearly in the photonic band gap frequency range. The dielectric micro pattern of the flat photonic crystal composed of the hexagonal acryl tablets was fabricated successfully as shown in figure 8. The nanometer sized titania particles were verified to disperse uniformly in the tablets, and the dielectric constant of the composite material was measured as 40. Figure 9-(a) and (b) shows transmission spectra obtained by the TDS and TLM methods. The measured forbidden gap agrees with the calculated photonic band gap in the frequency rage form 0.3 to 0.6 THz approximately. The maximum attenuation was measured as about -50 dB, and the minimum transmittance showed below 0.001 percent. Figure 10 shows the simulated distribution of electric field intensity in the micro pattern at the edge frequencies of the photonic band gap. The standing wave vibrations were localized and concentrated strongly in the air gap and dielectric tablets at the lower and higher band gap edges, respectively. The fabricated flat photonic crystals are considered to be applied to the novel sensors. The terahertz wave vibrations are well known to harmonize with molecule vibrations in various saccharides and proteins. The air path network can include the water solvents and localize the terahertz waves. Therefore, the characteristic spectra will be observed according to the biochemical material compositions and structures.

CONCLUSIONS

Micro patterns composed of polygon acryl tablets with titania particles dispersions were arranged periodically in two dimensions by using a micro patterning stereolithography. These flat photonic crystals were verified to be able to exhibit a forbidden band of opaque region in terahertz wave frequency ranges. In transmission spectra through square tablets arrangements, a localized mode of a transmission peak was clearly formed at a specific frequency, and electromagnetic energy concentration was verified in the periodic pattern. And terahertz wave energies were concentrated and localized in thin layers thorough the multiple resonations and standing wave formation. The fabricated micro pattern can include various types of solutions into their air gaps between the polygon tablets, therefore, it will be applied for novel micro sensors to detect useful biological materials.

ACKNOWLEDGMENTS

This study was supported by Priority Assistance for the Formation of Worldwide Renowned Centers of Research - The Global COE Program (Project: Center of Excellence for Advanced Structural and Functional Materials Design) from the Ministry of Education, Culture, Sports, Science and Technology (MEXT), Japan.

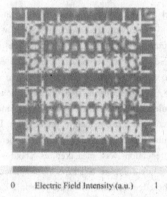

0   Electric Field Intensity (a.u.)   1

Figure 7   An intensity profile of electric field on a cross sectional plane in the dielectric micro pattern calculated at the peak frequency of the localized mode by using the TLM simulation of a finite difference time domain (FDTD) method.

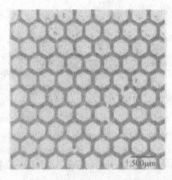

Figure 8   The micro pattern with the periodic arrangement of the dielectric tablets fabricated by using the micro stereolithography of computer aided designing and manufacturing. The titania particles were dispersed in to the hexagonal acrylic tablets.

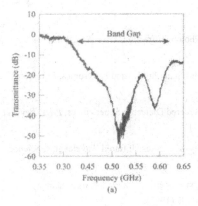 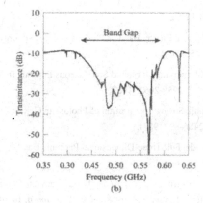

Figure 9   The transmission amplitudes of the terahertz wave through the dielectric micro pattern with the hexagonal tablets arrangement. The spectra (a) and (b) are measured and calculated properties by using the TDS and TLM methods, respectively.

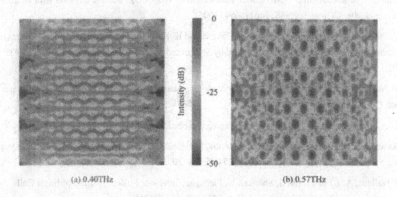

Figure 10   The intensity profiles of electric field on the cross sectional plane in the dielectric micro pattern. The energy maps (a) and (b) are calculated and constructed at the boundary frequencies of lower and higher band gap edges by using the TLM simulation, respectively.

REFERENCES

[1] K. Ohtaka, Energy Band of Photons and Low-Energy Photon Diffraction, *Phys. Rev. B*, **19**, 5057-5067 (1979).

[2] E. Yablonovitch, Inhibited Spontaneous Emisson in Solid-State Physics and Electronics, *Phys. Rev. Let.*, **58**, 2059-2062 (1987).

[3] S. John, Strong Localization of Photons in Certain Disordered Dielectric Superlattices, *Phys. Rev. Let.* **58**, 2486-89 (1987).

[4] S. Noda, Full Three-Dimensional Photonic Bandgap Crystals at Near-Infrared Wavelengths, *Science*, **289**, 604-6 (2000).

[5] R. E. Brown, D. C. Parker and E. Yablonovich, Radiation Properties of a Planar Antenna on a Photonic-Crystal Substrate, *J. Opt. Soc. Am. B*, **10**, 404-407 (1993).

[6] S. Noda, Three-Dimensional Photonic Crystals Operating at Optical Wavelength Region, *Physica B*, **279**, 142-49 (2000).

[7] S. Kirihara, M. Takeda, K. Sakoda and Y. Miyamoto, Control of Microwave Emission from Electromagnetic Crystals by Lattice Modifications, *Solid State Com.*, **124**, 135-139, (2002).

[8] S. Kanehira, S. Kirihara and Y. Miyamoto, Fabrication of TiO_2-SiO_2 Photonic Crystals with Diamond Structure, *J. Am. Ceram. Soc.*, **88**, 1461-1464 (2005).

[9] S. Kirihara, Y. Miyamoto, K. Takenaga, M. Takeda and K. Kajiyama, Fabrication of Electromagnetic Crystals with a Complete Diamond Structure by Stereolithography, *Solid State Com.*, **121**, 435-439 (2002).

[10] V. M. Van Exter, C. Fattinger and D. Grischkowsky, Terahertz Time-domain Spectroscopy of Water Vapor, *Opt. Let.*, **14**, 1128-1130 (1989).

[11] D. Clery, Brainstorming Their Way to an Imaging Revolution, *Science*, **297**, 761- 763 (2002).

[12] K. Kawase, Y. Ogawa, Y. Watanabe and H. Inoue, Non-destructive Terahertz Imaging of Illicit Drugs Using Spectral Fingerprints, *Opt. Exp.*, **11**, 2549-2554 (2003).

[13] V. P. Wallace, A. J. Fitzgerald, S. Shankar, N. Flanagan, Terahertz Pulsed Imaging of Basal Cell Carcinoma ex Vivo and in Vivo, *Brit. J. Der.*, **151**, 424–432 (2004).

[14] Y. Oyama, L. Zhen, T. Tanabe and M. Kagaya, Sub-Terahertz Imaging of Defects in Building Blocks, *NDT&E Int.*, **42**, 28-33 (2008).

[15] W. Chen, S. Kirihara and Y. Miyamoto, Fabrication and Measurement of Micro Three-Dimensional Photonic Crystals of SiO_2 Ceramic for Terahertz Wave Applications, J. Am. Ceram. Soc., **90**, 2078-2081 (2007).

[16] W. Chen, S. Kirihara and Y. Miyamoto, Three-dimensional Microphotonic Crystals of ZrO_2 Toughened Al_2O_3 for Terahertz Wave Applications, *Appl. Phys. Let.*, **91**, 153507-1-3 (2007).

[17] W. Chen, S. Kirihara and Y. Miyamoto, Fabrication of Three-Dimensional Micro Photonic Crystals of Resin-Incorporating TiO_2 Particles and their Terahertz Wave Properties, J. Am. Ceram. Soc., **90**, 92-96 (2007).

[18] W. Chen, S. Kirihara and Y. Miyamoto, Static Tuning Band Gaps of Three-dimensional Photonic Crystals in Subterahertz Frequencies, *Appl. Phys. Let.*, **92**, 183504-1-3 (2008).

[19] H. Kanaoka, S. Kirihara and Y. Miyamoto, Terahertz Wave Properties of Alumina Microphotonic Crystals with a Diamond Structure, *J. Mat. Res.*, **23**, 1036-1041 (2008).

[20] Y. Miyamoto, H. Kanaoka and S. Kirihara, Terahertz Wave Localization at a Three-dimensional Ceramic Fractal Cavity in Photonic Crystals, *J. Appl. Phys.*, **103**, 103106-1-5 (2008).

[21] S. Kirihara and Y. Miyamoto, Terahertz Wave Control Using Ceramic Photonic Crystals with Diamond Structure Including Plane Defects Fabricated by Micro-stereolithography, *Int. J. Appl. Ceram. Tech.*, **6**, 41-44 (2009).

[22] S. Kirihara, T. Niki and M Kaneko, Three-Dimensional Material Tectonics for Electromagnetic Wave Control by Using Micoro-Stereolithography, *Ferroelectrics*, **387**, 102-111 (2009).

INTERGRANULAR PROPERTIES AND STRUCTURAL FRACTAL ANALYSIS OF BaTiO₃-CERAMICS DOPED BY RARE EARTH ADDITIVES

V. V. Mitic[1, 2], V. Pavlovic[2], V. Paunovic[1], J. Purenovic[1], Lj. Kocic[1], S. Jankovic[2], I. Antolovic[1], D. Rancic[1]

[1]Faculty of Electronic Engineering, University of Nis, Nis, Serbia
[2]Institute of Technical Sciences, Serbian Academy of Sciences and Arts, Belgrade, Serbia

ABSTRACT

Ferroelectric BaTiO₃ as one of the most important ceramics materials in electronic, used on wide range of applications, can be modified with various dopant ions. In this paper, the influence of Er₂O₃, Yb₂O₃, Ho₂O₃ and La₂O₃, on microstructure, microelectronic and dielectric properties of BaTiO₃-ceramics has been investigated. The solid solubility of rare-earth ions in the BaTiO₃ perovskite structure has been studied by different methods of structural analysis including SEM-JEOL 5300 and energy dispersive spectrometer (EDS) system. BaTiO₃-ceramics doped with 0.01 up to 1 wt% of rare-earth additives were prepared by conventional solid state procedure and sintered from 1320°C to 1380°C for four hours. We also applied the fractal method in microstructure analysis of sintered ceramics, especially as influence on intergranular capacitor and dielectric peoperties of BaTiO₃-ceramics. This fractal nature effect has been used for better understanding integrated microelectronics characteristics and circuits.

INTRODUCTION

It is well known that, BaTiO₃-ceramics are one of the most important electronics materials, used in different applications: sensors, capacitors, multilayer ceramic capacitors, PTC and NTC resistors. Since grain size and distribution considerably affect electrical properties of these materials, correlation of their microstructure and electrical properties has been investigated most extensively by numerous authors [1-3].

It has been shown that electrical properties of undoped and doped BaTiO₃-ceramics are mainly controlled by barrier structure, domain motion of domain boundaries and the effects of internal stress in the grains. Therefore, microstructure properties of barium-titanate based materials, expressed in grain boundary contacts, are of basic importance for electric properties of these material [4, 5].

Since, both, intergranular structure and electrical properties, depend on ceramics diffusion process, it is essential to have an equivalent circuit model that provides a more realistic representation of the electrical properties. Recently, it has been established that modeling of random microstructures like aggregates of grains in polycrystals, patterns of intergranular cracks, and composites, theory of Iterated Function Systems (IFS) and the concept of Voronoi tessellation can be used [6]. Taking this into account, in this article we have developed methods for modeling grain geometry, grain boundary surface and geometry of grain contacts of doped BaTiO₃-ceramics. Most of these methods are based on BaTiO₃ microstructure analysis and on fractal correction which expresses the irregularity of grains surface through fractal dimension [7-9]. Also, we showed some results for intergranular contact surfaces based on mathematical statistical methods and calculations.

In this article, in order to enable reconstruction of microstructure configuration, grains shapes and intergranular contacts, electroceramics microstructure fractal analysis and characterization has been performed.

EXPERIMENTAL WORK

The samples were prepared from high purity (>99.98%) commercial BaTiO₃ powder (MURATA) with [Ba]/[Ti]=1,005 and reagent grade Er₂O₃, Yb₂O₃, Ho₂O₃ and La₂O₃ powders (Fluka chemika). Er₂O₃, Yb₂O₃, Ho₂O₃ and La₂O₃ dopants were used in the amount to have 0.01, 0.1, 0.5 and 1wt% Er, Yb, Ho and La in BaTiO₃. Starting powders were ball-milled in ethyl alcohol for 24 hours using polypropylene bottle and zirconia balls. After drying at 200°C for several hours, the powders were pressed into disk of 7·mm in diameter and 3·10⁻³m in thickness under 120 MPa. The compacts were sintered from 1320°C to 1380°C in air for four hours. The microstructures of sintered and chemically etched samples were observed by scanning electron microscope (JEOL-JSM 5300) equipped with energy dispersive X-ray analysis spectrometer (EDS-QX 2000S system). Prior to electrical measurements silver paste was applied on flat surfaces of specimens. Capacitance, dissipation factor and impedance measurements were done using Agilent 4284A precision LCR meter in the frequency range of 20 to 1MHz. The illustrations of the microstructure simulation, were generated by Mathematica 6.0 software.

RESULTS AND DISCUSSION

The SEM investigation has shown that the microstructures of samples doped with Er₂O₃, Yb₂O₃ and Ho₂O₃ exhibit similar features The samples sintered with Er₂O showed that the grains were irregularly polygonal shaped (Fig. 1a), although in Yb and Ho doped BaTiO₃ the grains are more spherical in shape (Fig. 2.b and Fig. 3.b).

For the lowest concentration of dopants, the size of the grains was large (up to 60 μm), but by increasing the dopant concentration the grain size decreased. As a result, for 0.5wt% of dopant the average grain size was from 7 to 10 μm, as can be seen from the cumulative grain size distribution curves for doped BaTiO₃, given in Fig. 5. For the samples doped with 1 wt% of dopant grain size drastically decreased to the value of only few μm. Spiral concentric grain growth which has been noticed for the samples sintered with 0.1 wt% of Er₂O₃ or Yb₂O₃ disappeared when the concentration increased up to 1 wt% of dopant. For these samples the formation of the "glassy phase" indicated that the sintering was done in liquid phase. The microstructure of samples doped with La is with lower grain size-from 5μm, for lower La concentration, up to 1 μm, for concentration 1wt% of La.

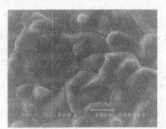 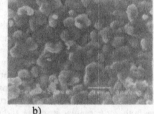

a)   b)

Fig. 1. SEM micrographs of doped BaTiO₃, sintered at 1320⁰C a)0.1 wt% Er₂O₃ and b)1.0 wt% Er₂O₃

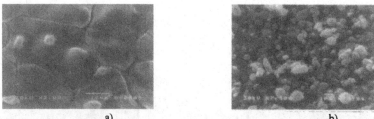

a)   b)

Fig. 2. SEM micrographs of doped BaTiO$_3$, sintered at 1320^{0}C a)0.1 wt% Yb$_2$O$_3$ and b)1.0 wt% Yb$_2$O$_3$

a)   b)

Fig. 3. SEM micrographs of doped BaTiO$_3$, sintered at 1320^{0}C a)0.1 wt% Ho$_2$O$_3$ and b)1.0 wt% Ho$_2$O$_3$

a)   b)

Fig. 4. SEM micrographs of doped BaTiO$_3$, sintered at 1320^{0}C a)0.1 wt% La$_2$O$_3$ and b)1.0 wt% La$_2$O$_3$

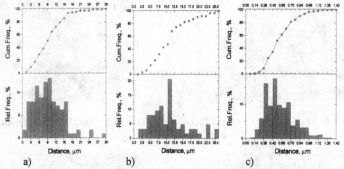

Fig. 5. Cumulative grain size distribution curves for doped $BaTiO_3$ with a) 0.5% Er_2O_3, b) 0.5% Yb_2O_3 and c) 0.5% Ho_2O_3

EDS analysis [10]. for all samples has shown, that for small concentrations of dopants the distribution is uniform, while the increase of dopant concentration led to the coprecipitation between grains .

Our new approach includes fractal geometry in describing complexity of the spatial distribution of electroceramic grains. The best fractal model is a sponge model or a kind of three –dimensional lacunary set (a set with voids). The structure of tetrahedral influence may be established in each spatial sense, which means that one has a tetrahedral lattice that fills the space.

FRACTAL NATURE STRUCTURE MODEL

Irregularity of the surface of ceramic grains can be expressed using the term called *fractal dimension* [11]. Being attached to some fractal set G, this number D=dim(G) strictly exceeds its topological dimension. Fractal dimension can be used in two ways: to improve the numerical value of the size of the contact area S, and to make a geometrical model of the grain that faithfully represents it. So the initial problem is to extract the fractal dimension from the measurement data. For this purpose we use the Richardson's law [11] that gives the relationship between the measure scale ε and the length $L(\varepsilon)$ of some irregular contour $L(\varepsilon)=K\,e^{1-D}$, where K is some positive constant and D is just the fractal dimension of this contour. Using the microphotography, we measure $L(\varepsilon)$ for different magnification scales ε. The data points $(Ln(L(\varepsilon_j)), Ln(\varepsilon_j))$ are approximated by a linear function using the least square method. The slope of this line is approximately $1-D$. By this technique we got the following dimensions for six different grains: $D_1 \approx 1.071890$, $D_2 \approx 1.071993$, $D_3 \approx 1.061486$, $D_4 \approx 1.0732408$, $D_5 \approx 1.116055$ and $D_6 \approx 1.0119546$. The average value $D \approx 1.0677699$ can be accepted as an approximation of the fractal dimension of the contour line of a $BaTiO_3$-ceramics grain

In order to develop model for the reconstruction of grains surface, theory of fractal sets has to be introduced.

A wide class of fractal sets can be defined as invariant under a contractive mapping. This overall definition leads to the theory of Iterated Function Systems (IFS) introduced by Barnsley [12].

Let $\{w_i\}_{i=1}^n$ be a set of contractive Lipschitz mappings with factors $|s_i| < 1$, $i = 1, ..., n$, of the metric space $(\mathbf{R}^r, d)$. Let $H(\mathbf{R}^r)$ be the set of all nonempty compact subsets of $\mathbf{R}^r$, and let h denote the Hausdorff metric induced by d, so that $(H(\mathbf{R}^r), h)$ is a complete metric space. Let $p_1, ..., p_n$ be positive real numbers such that $\Sigma p_i = 1$. The system

$$\omega = \left\{ R^\nu ; w_1,...,w_n; \ p_1,...,p_n \right\} \tag{1}$$

is called (*hyperbolic*) Iterated Function System, (or *IFS*) with probabilities. The associated Hutchinson operator $W : H(\mathbf{R'}) \rightarrow H(\mathbf{R'})$, defined by

$$W(G) = \bigcup_{i=1}^{n} w_i(G), \quad G \subset \mathbf{R}, \tag{2}$$

is a contraction of $(H(\mathbf{R'}), h)$ with the Lipschitz factor $s = \max_i |s_i| < 1$. So, there is a unique set F $H(\mathbf{R'})$ being a fixed point of W, i.e.

$$F = W(F) = \lim_{k \to \infty} W^{ok}(G), \ G \in H(\mathbf{R}), \tag{3}$$

which is called the *attractor of* ω, and denoted by att(ω). Sometimes it is also called *deterministic fractal*[3]. But even the simplest IFS codes, i.e. those made up of affine contractions, are not flexible enough. For example they are not affine invariant. Here, we use a stochastic approach to build an affine invariant code, called AIFS, containing only linear transformations.

Considering their unique properties, for grains surface modeling, method of Voronoi tessellation has been used. It is important to emphasize that although polycrystal modeling generally represents a 3D problem a 2D Voronoi tessellation approximation is used. This assumption is based on the fact that, we have more interest in the properties of the microstructure surface, rather than the inner structure.

A Voronoi tessellation represents a cell structure constructed from a Poisson point process by introducing planar cell walls perpendicular to lines connecting neighboring points. This results, in a set of convex polygons/polyhedra (Fig. 6), embedding the points and their domains of attraction, which partitioned the underlying space.

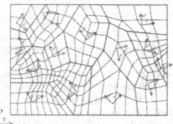

Fig. 6. Voronoi tessellation of 14-grain aggregate with grain boundaries, orientations of crystal lattices, finite element mesh and boundary conditions.

Generating set of points that should represent the distribution of micro-grains The idea is that points are distributed by a fractal law. In this case, the distribution is semi accident and implemented so that each point dislocates in relation to its position on the imaginary proper lattice. The main idea is to establish correlation between the different points-peaks and micro values between two contact surfaces, which are practically representing the new level, more complex and realistic, micro intergranular capacitors-impendances network (Fig. 7).

  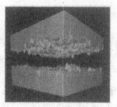

Fig. 7. Mapping Voronoi in 3D model

Together with this Voronoi model we combined the fractals, also. This kind of fractals sometimes is identified with deterministic constructions like *Cantor set*, *Sierpinski triangle* or *Sierpinski square*, with *Sierpinski pyramid* or *Menger sponge*, and so on

 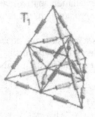 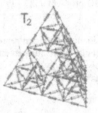

Fig. 8. The model of impedances between clusters of ceramics grains

The *Sierpinski pyramid* (Fig. 8) might be a quite adequate paradigm for the first instance inquiry. The starting pyramid T_0 and the first two iterations, give initial part of the orbit of so called Hutchinson operator W, that is $T_1 = W(T_0)$, and $T_2 = W(T_1) = W^2(T_0)$. The limit case, $T = W(T_0)$ is an exact fractal set with Hausdorff dimension $D_H = 2$.

The real morphology of BaTiO₃-ceramic grains makes this approximation reasonable. The volume of the sample is approximated by the cube of the following dimensions: 10 μm × 10 μm × 10 μm. Inside the cube spherical grains are distributed according to the log-normal distribution

$$f(l) = \frac{1}{\sigma \cdot l \sqrt{2\pi}} e^{-\frac{(\ln l - \ln \bar{l})^2}{2\sigma^2}} \tag{4}$$

where $\bar{l}$ is the middle grain size and $\sigma°$- is a dispersion. Values of these sizes for sintered BaTiO₃ (Fig. 9 and 10) are $\bar{l} = 0.73$ and $\sigma = 0.46$. Grains have the shape of polyhedron with a great number of flat surfaces N, so that N gravitates toward very great number approximate the spherical shape. The rate of grain penetration is estimated to 10-15% of the grain radius size. Searching through contacts and calculating their values are carried out by layers of $z=0.05$ size. By application of the corresponding computer program for grains contact identification and calculation, we obtained the distribution of the contact surfaces numbers through contact surfaces values. These curves are shown on Fig. 10a. and b.

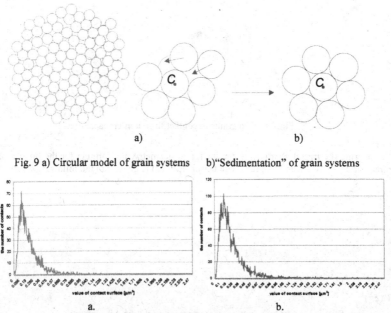

Fig. 9 a) Circular model of grain systems   b)"Sedimentation" of grain systems

a.   b.

Fig. 10. Diagrams showing the contact surfaces values obtained for the system with: a. 500 and b. 700 grains.

Diagram in Fig. 10a. refers to the case when 500 grains are in the cube and the maximum of the curve is for the contact surface value of 0.19 μm^2. For the curve shown on Fig. 10b. this value is approximately 0.22 μm^2 which is very close to the previous value. By summing up, all the contact values, through all intersection levels, the integral contact surface is calculated: 462.28 μm^2 for the case of 500 grains and 763.21 μm^2 for 700 grains inside the materials' volume used. It should be emphasized that these values should be taken as values of the integral surface on the model level

Due to diffusional forces that appear in sintering process we are ready to believe that an approximate form of a contact surface is the shape of a minimal surface - the surface with minimal area size. But, the microstructure of the material makes this surface to be fractal locally (Fig. 11).

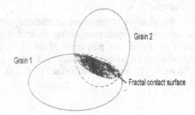

Fig. 11. An intergrain contact surface has fractal form.

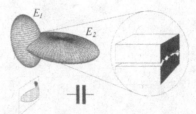

Fig. 12. Two grains in contact form a microcapacitor.

Considering the fractal approach to the intergrain geometry, the formula for the microcapacity of intergranular condensor seen as a planar condensor (Fig. 12) is given into the following form

$$C' = \varepsilon_0 \varepsilon_B' \frac{S_c}{x},$$ (5)

Where

$$\varepsilon_B' = \alpha \varepsilon_B,$$

and

$$\alpha = (N\xi^2)^k,$$

is a correction factor obtained by a constructive approach to the fractal surface

$$C = \varepsilon_0 \varepsilon_B \alpha \cdot \frac{S}{x} = \varepsilon_0 \varepsilon_B (N\xi^2)^k \cdot \frac{S}{x},$$ (6)

where ε_0, ε_B are dielectric constants in vacuum and in BaTiO₃-ceramics material respectively; S - the area of the 'plates' and x - distance between condensor 'plates', i.e. the condensor thickness and $\alpha = (N\xi^2)^k$, is a correction factor obtained by a constructive approach to the fractal surface.

This approach uses an iterative algorithm that iterates N self-affine mappings with a constant contractive (Lipschitz) factor $|\xi|<1$ k-times[3]. The underlying theory and techniques for choosing the appropriate mappings are given in previous work. Typically, $a = D - D_T$, where $D \approx 2.08744$ is the fractal (Hausdorff) dimension of intergrain contact surface and $D_T = 2$ is topological dimension of the surface. As it is found, BaTiO₃-ceramics contact surfaces are of *low-irregularity* which is characterized by the small difference $D - D_T \approx 0.08744$.

Derived formula (6) indicates the increase of the value of microcapacity when fractal approach is applied. Thus, more accurate calculation of microcapacitance generated in grains contact can be carried out leading to a more exact estimation of dielectric properties of the whole sample.

Taking this into account, calculations on microcapacitance generated in grains contacts of BaTiO₃ doped with 0.5 wt.% Er and 0.5% wt.% Yb have been performed (Fig. 13).

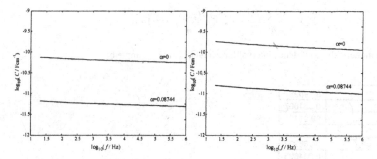

Fig. 13. Microcapacitance vs. frequency for BaTiO₃ doped with a) 0.5 wt.% Er b) 0.5% wt.% Yb

In order to determine an algebaric equation describing equivalent intergranular impedance in terms of circuit parameters following equation can be used:

$$Z(s) = \frac{1 + CR \cdot s + CL \cdot s^2}{(Cp + C) \cdot s + CpCR \cdot s^2 + CpCL \cdot s^3}$$

where $s = j\omega$, $\omega = 2\pi f$, f is frequency. Each set of defined values for model parameters results in corresponding frequency characteristics. In order to bring our model closer to real physics of intergranular processes as well as to apply the model on different grain's structures, we defined five different sets of model parameters (C1, C, L, R).

As we mentioned before, here is presented a statistical approach to the investigation of BaTiO₃-ceramic grains, concerning the relationship between the capacitance and the area of the contact surface. The method applies to all kinds of grains, to every ceramic structure, regardless of the actual shape of the ceramic grains and of the actual shape of the contact area between the grains. The only thing that matters is the area of the contact surface on the squares that we investigate. We assume that the distribution of the contact area is the same on each square, with the unknown mean θ and the variance σ^2.

We perform the following experiment n times, that is, we choose randomly n squares with prescribed vertices on BaTiO₃-ceramic electronic material that was subject to sintering, and we measure the areas of contact surfaces on these n squares. As the result (Table 1.), we obtain the values $X_1, X_2, ..., X_n$ - the measure of contact areas on each of the corresponding squares. From the sample we obtain the sample mean $\overline{X} = \frac{1}{n}\sum_{i=1}^{n} X_i$, the sample variance $\overline{\sigma}^2 = \frac{1}{n}\sum_{i=1}^{n}(X_i - \overline{X})^2$ and the unbiased estimate of the variance

$$\overline{\sigma}_0^2 = \frac{1}{n-1}\sum_{i=1}^{n}(X_i - \overline{X})^2.$$

For chosen $a = 0.05$, the confidence interval for the unknown mean μ is the following one:

$$(\overline{X} - \frac{t_{n-1,\alpha/2}\overline{C}}{n^{1/2}} < \mu < \overline{X} + \frac{t_{n-1,\alpha/2}\overline{C}}{n^{1/2}}),$$

that is

$$P(\overline{X} - \frac{t_{n-1,\alpha/2}\overline{C}}{n^{1/2}} < \mu < \overline{X} + \frac{t_{n-1,\alpha/2}\overline{C}}{n^{1/2}}) = 1 - \alpha.$$

Table 1. Calculations of the confidence interval for contact surfaces (sample of the size 30)

Mean $\overline{X}$	99,90987
Standard Error	0,034007
Median	99,95791
Mode	100
Standard Deviation $\overline{C}$	0,186262
Sample Variance	0,034693
Kurtosis	21,94624
Skewness	-4,41491
Range	1,010101
Minimum	98,9899
Maximum	100
Count	30
Confidence Level(95,0%)	0,069551

$P(99.84 < \mu < 99.98) = 0.95$

Statistical calculations are done for all samples. As an example, mentioned parameters are given in Table 2. for sample with 0.1wt% Er$_2$O$_3$, sintered at 1320°C.

Table 2. Various statistical parameters for contact surfaces

0.1wt%Er$_2$O$_3$, 1320°C	
Mean	99,90987
Standard Error	0,034007
Median	99,95791
Mode	100
Standard Deviation	0,186262
Sample Variance	0,034693
Kurtosis	21,94624
Skewness	-4,41491
Range	1,010101
Minimum	98,9899
Maximum	100
Count	30
Confidence Level(95,0%)	0,069551

We assume that the relationship between the capacitance C and the contact area S is of the form $C = 6S$, where 6 is the unknown parameter that we want to estimate by taking the sample mean $\overline{X}$ instead of S in the formula $C = 6S$.

As a continuation in our investigation, for better and deeper characterization and understanding the ceramics material microstructure, we applied the methods which are recognized the fractal nature structure, and also Voronoi model and mathematical statistics calculations, in the purpose to establish multidimensional view on real ceramics structure. All of these analyses and different characterization methods are very important for prognosis the final, especially electric and ferroelectric BaTiO₃-ceramics properties. So, we began development the original intergranular impedance model based on intergranular capacity.

CONCLUSION

In this paper fractal geometry has been used to describe complexity of the spatial distribution of electroceramics grains. It has been shown that the control of shapes and numbers of contact surfaces on the level of the entire electroceramic sample, and over structural properties of these ceramics can be done. Also, the Voronoi model practically provided possibility to control the ceramics microstructure fractal nature. The mathematical statistic methods gave as possibility to establish the real model of prognosis-correlation: synthesis-structures-properties. The complex model of intergranular impedance is established using the equivalent electrical scheme characterized by corresponding frequency characteristic. According to the electroceramic microstructures the global impedance of the samples, which contains both resistor and capacity component, has been presented as a "sum" of many clusters of micro-resistors and micro-capacitors connected in tetrahedral lattice. All of these results give us multidimensional possibility to establish new approach for microelectronics integrated circuits in ceramics structures, as the method on the way of future microelectronics miniaturization.

ACKNOWLEDGMENTS

This research is a part of the project „Directed synthesis, structure and properties of multifunctional materials" (172057). The authors gratefully acknowledge the financial support of Serbian Ministry for Science and Technological Development for this work.

REFERENCES

1. V.V. Mitić, I. Mitrović, D. Mančić, "The Effect of CaZrO₃ Additive on Properties of BaTiO₃-Ceramics", Sci. Sint., Vol. 32 (3), pp. 141-147, 2000
2. P.W.Rehrig, S.Park, S.Trolier-McKinstry, G.L.Messing, B.Jones, T.Shrout Piezoelectric properties of zirconium-doped barium titanate single crystals grown by templated grain growth J. Appl. Phys. Vol 86 3, (1999) 1657-1661
3. V.P.Pavlovic, M.V.Nikolic, V.B.Pavlovic, N. Labus, Lj. Zivkovic, B.D.Stojanovic, Correlation between densification rate and microstructure evolution of mechanically activated BaTiO₃, Ferroelectrics 319 (2005) 75-85
4. D. Lu, X. Sun, M. Toda Electron Spin Resonance Investigations and Compensation Mechanism of Europium-Doped Barium Titanate Ceramics Japanese Journal of Applied Physics Vol. 45, No. 11, 2006, pp. 8782-8788
5. N.Hirose, A.West Impedance Spectroscopy of undoped BaTiO₃ Ceramics J.Am.Ceram.Soc, 79 [6] 1633-41 (1996)
6. V.V. Mitić, Lj. M. Kocić, I. Mitrović, A Model for Calculating Contact Surfaces of BaTiO₃-Ceramics - *Science of Sintering*, Vol. 30 (1), pp. 97-104, 1998
7. V.V.Mitic, V.B.Pavlovic, Lj. Kocić,V.Paunovic, D.Mancic, *Application of the Intergranular Impedance Model in Correlating Microstructure and Electrical Properties of Doped BaTiO3*, Science of sintering, (Vol. 41, No. 3 Page 247-256) 2009

8. Vojislav Mitić, V. Paunović, D. Mančić, Lj. Kocić, and Lj. Zivković and V.B. Pavlović, "Dielectric Properties of BaTiO$_3$ Doped with Er$_2$O$_3$, Yb$_2$O$_3$ Based on Intergranular Contacts Mode"l, Published in:*Ceramic Transactions Vol. 204; Advances in Electroceramic Materials,* pp. 137-144, 2009
9. V. Mitić, V.Pavlović, Lj.Kocić, V. Paunović. Lj.Živković, "Fractal geometry and properties of doped BaTiO$_3$ ceramics", Advances in Sciency and Technology, Vol 67, pp. 42-48, 2010
10. V.V.Mitić, V.B.Pavlović, M.Miljković, V.V.Paunović, B.Jordović, Lj.M.Zivković, "SEM and EDS analysis of BaTiO$_3$ doped with Yb$_2$O$_3$ and Ho$_2$O$_3$", 14th Europran Microscopy Congress, EMC 2008, Germany , pp.555-556, 2008
11. . Lj. M. Kocić, V.V. Mitić, M.M. Ristić, *Stereological Models Simulation of BaTiO3-Ceramics Grains* - Journal of Materials Synthesis and Processing, Vol. 6, No. 5, pp. 339-344 (1998)
12. Barnsley, M. F., *Fractals Everywhere,* Academic Press, 1993

DEVELOPMENT OF NUMERICAL METHOD FOR EVALUATING MICROSTRUCTURAL FRACTURE IN SMART MATERIALS

Hisashi Serizawa
Joining and Welding Research Institute, Osaka University
11-1 Mihogaoka, Ibaraki, Osaka 567-0047, Japan

Tsuyoshi Hajima and Seigo Tomiyama
Graduate School of Engineering, Osaka University
2-1 Yamadaoka, Suita, Osaka 565-0871, Japan

Hidekazu Murakawa
Joining and Welding Research Institute, Osaka University
11-1 Mihogaoka, Ibaraki, Osaka 567-0047, Japan

ABSTRACT
    In order to examine the microstructural behavior in the advanced multifunctional materials, a numerical method has been developed, where the elastic-plastic deformation of material was modeled by the ordinary finite element and the interface element was employed for simulating the interfacial behavior such as debonding & slipping at grain boundary and growth & propagation of crack. As a result of serial computations for the fracture behavior in an elastic perfectly plastic plate with a center crack, it was found that by changing the yield stress the fracture behavior could be divided into three modes, which were "Plastic Deformation Dominant Mode", "Transient Mode" and "Crack Growth Dominant Mode". Also, the fracture behavior in a two-dimensional virtual polycrystalline structure with an initial crack at grain boundary could be demonstrated by using this numerical method. Then, it can be concluded that the interface element would have a good potential to examine the elastic-plastic fracture behavior.

INTRODUCTION
    Advanced multifunctional materials have been developed by controlling their microstructure precisely or by joining various dissimilar materials. As for the practical use of these materials, their fracture behavior should be estimated theoretically, where not only elastic-plastic mechanical deformation but also debonding & slipping at grain boundary have to be taken into account. Although the finite element method (FEM) has been widely employed to estimate the fracture behavior, only elastic-plastic deformations were mainly discussed in the most of previous studies and the various types of criteria for the fracture have been examined through the stress-strain distributions computed and the experimental trials & errors.
    In order to develop the mechanical performance of these multifunctional materials, both strength and toughness are the primary properties to be achieved. The strength is for the performance against the plastic collapse, while the toughness is that against the failure accompanying the formation and growth of the crack. The former is represented by the yield stress, and the latter is described by the fracture toughness parameters such as K, G, J and CTOD. Comparing the strength and the toughness, the strength is a relatively easy concept to understand. The clear difference between them is that the toughness is connected to the mode of failure accompanying the formation and growth of crack. Thus, to study the toughness of materials, a mechanical model which directly represents such behavior of crack is necessary. As one of the methods to demonstrate the interfacial behavior and the behavior of crack in the finite element analysis, the interface element has been developed and the crack propagation behavior based on the classic fracture mechanics could be demonstrated where only the elastic deformation was considered [1-4].

In this research, in order to examine the applicability of interface element for the elastic-plastic behavior, the crack propagation behavior in a plate with a center crack was simulated by using the interface element. Also, by using a two-dimensional virtual polycrystalline model obtained through Voronoi tessellations, the influence of microstructure on the elastic-plastic deformation of grain and the debonding & slipping at grain boundary was studied by using the interface element.

INTERFACE ELEMENT

Essentially, the interface element is the distributed nonlinear spring existing between surfaces forming the interface or the potential crack surfaces as shown by Fig. 1. The relation between the opening of the interface δ and the bonding stress σ is shown in Fig. 2. When the opening δ is small, the bonding between two surfaces is maintained. As the opening δ increases, the bonding stress σ increases till it becomes the maximum value σ_{cr}. With further increase of δ, the bonding strength is rapidly lost and the surfaces are considered to be separated completely. Such interaction between the surfaces can be described by the interface potential. There are rather wide choices for such potential. The authors employed the Lennard-Jones type potential because it explicitly involves the surface energy γ which is necessary to form new surfaces. Thus, the surface potential per unit surface area ϕ can be defined by the following equation.

$$\phi(\dot{\varepsilon}_n, \dot{\varepsilon}_t) \equiv \phi_a(\dot{\varepsilon}_n, \dot{\varepsilon}_t) - \phi_b(\dot{\varepsilon}_n) \tag{1}$$

$$\phi_a(\delta_n, \delta_t) = 2\gamma \cdot \left\{ \left(\frac{r_0}{r_0 + \delta} \right)^{2N} - 2 \cdot \left(\frac{r_0}{r_0 + \delta} \right)^{N} \right\}, \quad \delta = \sqrt{\delta_n^2 + A \cdot \delta_t^2} \tag{2}$$

$$\phi_b(\delta_n) = \begin{cases} \dfrac{1}{2} \cdot K \cdot \delta_n^2 & (\delta_n \le 0) \\ 0 & (\delta_n \ge 0) \end{cases} \tag{3}$$

Where, δ_n and δ_t are the opening and shear deformation at the interface, respectively. The constants γ, r_0, and N are the surface energy per unit area, the scale parameter and the shape parameter of the potential function. In order to prevent overlapping in the opening direction due to a numerical error

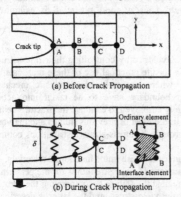

(a) Before Crack Propagation

(b) During Crack Propagation

Fig. 1   Representation of crack growth using interface element.

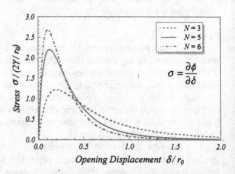

Fig. 2   Relationship between crack opening displacement and bonding stress.

in the computation, the second term in Eq. (1) was introduced and K was set to have a large value as a constant. Also, to model an interaction between the opening and the shear deformations, a constant value A was employed in Eq. (2). From the above equations, the maximum bonding stress, σ_{cr}, under only the opening deformation δ_n and the maximum shear stress, τ_{cr}, under only the shear deformation δ_t are calculated as follows.

$$\sigma_{cr} = \frac{4\gamma N}{r_0} \cdot \left\{ \left(\frac{N+1}{2N+1} \right)^{\frac{N+1}{N}} - \left(\frac{N+1}{2N+1} \right)^{\frac{2N+1}{N}} \right\} \tag{4}$$

$$\tau_{cr} = \frac{4\gamma N \sqrt{A}}{r_0} \cdot \left\{ \left(\frac{N+1}{2N+1} \right)^{\frac{N+1}{N}} - \left(\frac{N+1}{2N+1} \right)^{\frac{2N+1}{N}} \right\} \tag{5}$$

By arranging such interface elements along the crack propagation path as shown in Fig. 1, the growth of the crack under the applied load can be analyzed in a natural manner. In this case, the decision on the crack growth based on the comparison between the driving force and the resistance as in the conventional methods is avoided.

From the results of our previous researches using the interface elements, it was found that the failure mode and the stability limit depend on the combination of the deformability of the ordinary element in FEM and the mechanical properties of the interface element as controlled by the surface energy γ, the scale parameter r_0 and the interaction parameter A in Eq. (2); furthermore, the fracture strength in the failure problems of various structures might be quantitatively predicted by selecting the appropriate values for the surface energy γ, the scale parameter r_0 and the interaction constant A [1-4].

FRACTURE BEHAVIOR IN PLATE WITH A CENTER CRACK
Model for Analysis

Figure 3 shows a model for examining the applicability of interface element for elastic-plastic deformation, where the 100 mm square plate was loaded through the vertical displacement given on both ends of the plate. A length of center crack was assumed to 20 mm. According to the symmetry the model, a quarter of the model was employed for the finite element analysis as shown in Fig. 4, where the calculations were conducted as a two dimensional plain strain problem. In this analysis,

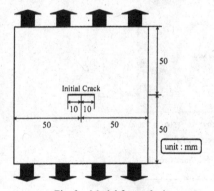

Fig. 3   Model for analysis.

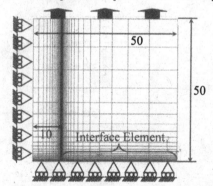

Fig. 4   Finite element mesh model for analysis.

the material properties of model were assumed to those of general steel since the most of them are well known as the results of previous researches [1,5]. So, Young's modulus and Poisson's ratio were assumed to 200 GPa and 0.3, respectively. In order to demonstrate the crack propagation behavior from the tip of center crack, the interface elements were arranged along the bottom of FEM model as shown in Fig. 4. Although the shear deformation would be little in this analysis, the functions of Eqs. (1) – (3) were employed as the interface potential. The surface energy γ, scale parameter r_0 and shape parameter N in the interface potential were assumed to 2.0 N/m, 1.0 m and 4, respectively according to the analysis of brittle fracture in the steel at low temperature [1,5], while the interaction parameter A in Eq. (2) was assumed to 1.0. Also, the mesh division near the crack tip was set to fine as shown in Fig. 4 since the mesh division near the crack tip would affect the stress-strain distributions computed by FEM.

Results and Discussions

In order to examine the effect of initial yield stress on fracture behavior of the plate with a center crack, the mechanical property was assumed to be elastic perfectly plastic which means elastic-plastic without any strain hardening. By changing the yield stress, the fracture load was computed and summarized into Fig. 5, where the maximum load calculated or the load at the loss of static equilibrium was defined as the fracture load in this research. The result of a plate without a center crack was also plotted as the case of "No Crack Growth" in this figure. In addition to these curves, the fracture load of an elastic plate with a center crack was drawn as the result of "Brittle Fracture Load". As it may be seen in Fig. 5, when the yield stress was small, the fracture load monotonically increased with increasing the yield stress and the curves for elastic perfectly plastic material with and without a center crack coincided. In this region, the plastic deformation near the crack tip expanded and the whole of plate changed to be plastic without any crack propagation. So, this area can be denoted as "Plastic Deformation Dominant Mode". On the other hand, when the yield stress was much larger, the fracture load approached the brittle fracture load of elastic plate and the crack propagated without plastic deformation. Namely, this region can be described as "Crack Growth Dominant Mode". Then, the middle area can be denoted as "Transient Mode" and the crack

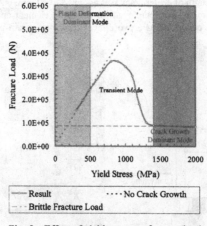

Fig. 5   Effect of yield stress on fracture load.

extension occurred with expanding plastic deformation near the crack tip. Since such crack propagation behavior is observed in the general metal materials, it was found that the elastic-plastic deformation of metal material could be demonstrated by using the interface element with assuming an appropriate value for the initial yield stress. Moreover, based on the difference between the strength and the toughness, it was revealed that there would be an optimum value of yield stress which could give the maximum fracture toughness. Although the maximum value of fracture load in Fig. 5 might change with the size of structures and existing cracks, it can be concluded that the interface element would have a good potential to examine the elastic-plastic fracture behavior.

EFFECT OF MICROSTRUCTURE ON FRACTURE BEHAVIOR
Model for Analysis
In order to study the influence of microstructure on fracture behavior, a two-dimensional virtual polycrystalline model of 20 grains was produced in 1 mm square through Voronoi tessellations as shown in Fig. 6. Based on this model, finite element model including the interface elements was made, where all grains were divided by almost uniform ordinary finite elements, which was about 10 m square, and the interface elements were arranged along all grain boundaries. Total number of elements was 164,484. Although the mechanical property of practical grain is anisotropic due to its crystal orientation, the property in this model was assumed to be isotropic as a preliminary research. Namely, Young's modulus and Poisson's ratio were set to 200 GPa and 0.3, respectively. The surface energy γ and the shape parameter N of interface element were assumed to 2.0 N/m and 4 as same as the previous study for the plate with a center crack. Since the shear strength at grain boundary is considered to be smaller than the bonding strength, the interaction parameter A in Eq. (2) was set to 0.25, which means that the maximum shear strength is half of the maximum bonding strength according to Eq. (4). The fracture behavior of virtual polycrystalline model was examined as a plain strain problem by applying the vertical displacement given on upper end of model, where an initial crack was assumed at one of grain boundaries. Especially, the influences of initial yield stress in grain and bonding strength at grain boundary on the stress-strain relationship were studied assuming elastic perfectly plastic deformation in grain.
Results and Discussions

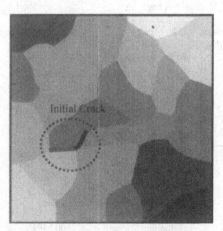

Fig. 6   Virtual polycrystalline model.

As an initial crack at grain boundary, one of three types which were "Short", "Medium" and "Long" was set as shown in Fig. 6. In order to examine the effect of bonding strength at grain boundary on the fracture behavior, the bonding strength was changed to 1500, 2000 and 2500 MPa assuming the initial yield stress in grain as 780 MPa. The computed stress-strain curves were summarized into Fig. 7. From this figure, it was found that the total strain increased with increasing the bonding strength regardless of the initial crack length. The reason of this behavior can be considered that the growth of resistance to crack propagation due to the increment of bonding strength would lead to the larger deformation of grains. Also, in the cases of "Long", it was revealed that the initial crack was easy to propagate although the grain near the crack tip achieved to the yield stress, and then the maximum stress of polycrystalline model could not achieve to the yield stress of grain.

On the other hand, the yield stress was changed to 580 and 1000 MPa in the condition that the bonding strength was set to 2000 MPa, in order to evaluate the influence of yield stress in grain on the fracture behavior. The stress-strain relationships obtained were summarized into Fig. 8. From Figs. 7(b) and 8, it was found that although the maximum stress of polycrystalline model increased with increasing the yield stress, the total strain decreased regardless of the initial crack length. Since the stress near the crack tip achieved to the yield stress, the difference between the stress near the crack tip

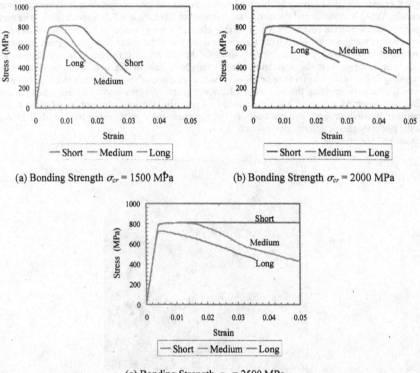

(a) Bonding Strength σ_{cr} = 1500 MPa

(b) Bonding Strength σ_{cr} = 2000 MPa

(c) Bonding Strength σ_{cr} = 2500 MPa

Fig. 7   Effect of bonding strength on fracture behavior (Yield Stress = 780 MPa).

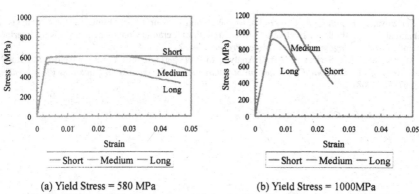

(a) Yield Stress = 580 MPa   (b) Yield Stress = 1000MPa

Fig. 8   Effect of yield stress on fracture behavior (Bonding Strength σ_{cr} = 2000 MPa).

and the bonding stress would become smaller with increasing the yield stress. So, the reason of this reduction in total strain can be considered that the initial crack would be easy to extend in the cases with larger yield stress and the total deformation of grain would become small.

Although these computations for the polycrystalline model were very simple examinations, it can be concluded that FEM with the interface element would be applied for studying the microstructural fracture behavior with considering both elastic-plastic deformation of grain and debonding & slipping at grain boundary.

CONCLUSIONS

As a tool for examining the microstructural fracture behavior in the multifunctional materials, a finite element method with the interface element has been developed and its potentiality was studied through the simple two problems. The conclusions can be summarized as follows.

(1) As a result of serial computations for the fracture behavior in an elastic perfectly plastic plate with a center crack, it was found that by changing the yield stress the fracture behavior could be divided into three modes, which were "Plastic Deformation Dominant Mode", "Transient Mode" and "Crack Growth Dominant Mode".

(2) The fracture behavior in a two-dimensional virtual polycrystalline structure with an initial crack at grain boundary could be demonstrated through the modeling of grains by the ordinary finite elements and of grain boundaries by the interface elements.

(3) The interface element would have a good potential to examine the elastic-plastic fracture behavior.

REFERENCES

[1] H. Murakawa, H. Serizawa and Z.Q. Wu, "Computational Analysis of Crack Growth in Composite Materials Using Lennard-Jones Type Potential Function," *Ceramic Engineering and Science Proceedings*, **20** [3], 309-316 (1999).

[2] H. Serizawa, H. Murakawa and C.A. Lewinsohn, "Modeling of Fracture Strength of SiC/SiC Composite Joints by Using Interface Elements," *Ceramic Transactions*, **144**, 335-342 (2002).

[3] H. Serizawa, C. A. Lewinsohn, M. Singh and H. Murakawa, "Evaluation of Fracture Behavior of Ceramic Composite Joints by Using a New Interface Potential", *Materials Science Forum*, **502**, 69-74 (2005).

[4] H. Serizawa, D. Fujita, C. A. Lewinsohn, M. Singh and H. Murakawa, "Finite Element Analysis of Mechanical Test Methods for Evaluating Shear Strength of Ceramic Composite Joints Using Interface Element", *Ceramic Engineering and Science Proceedings*, **27** [2], 115-124 (2006).
[5] H. Murakawa and H. Serizawa, "Arrest Analysis of Propagating Brittle Crack in Welded Structure", *Proceedings of International Symposium on Structures under Earthquake, Impact, and Blast Loading 2008*, 67-71 (2008).

FABRICATION OF CERAMIC DENTAL CROWNS BY USING STEREOLITHOGRAPHY AND POWDER SINTERING PROCESS

Satoko Tasaki and Soshu Kirihara
Joining and Welding Research Institute, Osaka University
11-1, Mihogaoka, Ibaraki, Osaka

Taiji Soumura
School of Dentistry, Osaka University
1-8, Yamadagaoka, Suita, Osaka

ABSTRACT

Fabrication processes of ceramic dental crown with low risks for metallic allergies and aesthetic sensuousness for natural human teeth are investigated and developed actively in worldwide medical industries. In this investigation, the dental crown models of acrylic resins including ceramic particles were fabricated by using laser scanning stereolithography. Moreover, complete ceramic objects as biomedical components were created successfully through powder sintering processes. Graphic data obtained by computer tomography scanning were converted into file sets of cross sectional images through slicing operations. Subsequently, photo sensitive acrylic resins including alumina at 60 vol. % were spread on a substrate with 60 μm in layer thickness. An ultraviolet laser beam of 100 μm in spot size was scanned on the slurry surface to create cross sectional images. After these automatic micro stacking processes, the dental crown models were fabricated. These precursors were dewaxed at 700 °C and sintered at 1400 to 1600 °C in air atmosphere, and uniformly dense and defect free microstructures were obtained successfully. Through bending tests for plate specimens, these ceramic bodies obtained by the optimized heat treatment could exhibit enough intensities required for the single crown in present dental technologies.

INTRODUCTION

All ceramic dental crowns to cover and protect human teeth or dental implant have been investigated and developed actively in world wide medical industries in order to realize advantages of aesthetic sensuousness and to avoid serious risks for metallic allergies. Various freeform fabrication techniques of ceramic crowns were invented and used.[1-3] In this investigation, the dental crown models of acrylic resins including alumina particles were fabricated by using laser scanning stereolithography of a computer aided design and manufacturing (CAD/CAM). According to graphic data of the dental crowns obtained by using a computer tomography scanning, the dense objects of alumina ceramics as biomedical components were fabricated successfully through powder sintering processes.[4-5] Moreover, the formed crowns were coated by dental glasses to improve aesthetic and mechanical properties. Part

141

accuracies, ceramic microstructures and mechanical properties were discussed for the fabricated dental crown models. The possibilities for the dental applications will be considered.

EXPERIMENTAL PROCEDURE

The dental crown models and flexural test specimens of 4.5×22×1.0 mm in dimensions were fabricated by using the stereolithography. Figure 1 shows schematic illustrations of the fabrication system. Three dimensional graphic models designed through the CAD software were converted into the stereolithography file format and sliced into a series of two dimensional plane data with uniform thickness. These numerical data were transferred into the process equipment (D-MEC, SCS-300P). Slurry material was prepared through mixing alumina particles (Showa Denko, AL-170) with photo sensitive acryl resin at 60 to 70 volume %. The mixed paste was spread and smoothed on a flat substrate. An ultra violet laser beam of 355 nm in wave length was scanned on the deposited layer to create cross sectional planes. After these layer stacking processes, solid components were fabricated. These precursors were de-waxed at 600 to 800 °C for 2 hs and sintered at 1400 to 1600 °C for 2 hs in the air atmosphere. The sintered ceramic components were coated with La_2O_3-B_2O_3-Al_2O_3-SiO_2 dental glass (Zahnfabrik, In-Ceram-Alumina) through heat-treatment at 1100°C for 2 hs. The ceramic microstructures were observed by using a scanning electron microscope. The flexural strengths of plate specimens were measured by using a three point bending test machine.

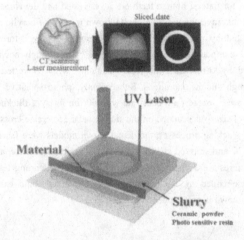

Fig. 1   Schematic illustrated data processing, laser scanning and layer stacking in stereolithography technique of a computer aided design and manufacturing (CAD/CAM) procedure.

RESULTS AND DISCUSSION

The stereolithographic composite model and the sintered alumina crown are shown in Figure 2. This model de-waxed at 600 °C for 2 hs with the heating rate of 0.5 °C/min in the air, and sintered at 1600 °C for 2 hs with the heating rate of 8 °C/min in the air. The average liner shrinkages of sintered samples were 7 % for the horizontal direction and 9 % for the vertical. The relative density of sintered alumina crown was 97 %. Macroscopic damages and deformations were not observed.

Cross sectional microstructures of sintered bodies and glass infiltrate samples are shown in Figure 3. The relative densities of sintered bodies were 79, 83 and 92 % after sintering at 1400, 1500 and 1600 °C, respectively. After glass infiltration, the relative densities of sample were 98, 99 and 99 % sintered at 1400, 1500 and 1600 °C, respectively. The large cracks propagate parallel to the stacked layers formed in the stereolithographic processes. Through the glass infiltrate treatments, these cracks became inconspicuous. The maximum flexural strength was achieved 587±91 MPa by sintering at 1500 °C with the glass infiltration. This mechanical property is acceptable level for the dental using.

The bending strength exhibits numerical fluctuations of about 200 MPa in maximum amplitude, due to macro crack formations in the powder sintering process. To increase the mechanical strength and reliability of the alumina dental crowns, the different types of photo sensitive resins were mixed to realize gradually gasification. The resin gasification occurred rapidly at two different temperature regions form 300 to 400 °C and 600 to 700 °C. Figure 4 shows the optimized de-waxing treatment pattern according to the measured mass reduction rate of the mixed resins. After the precisely controlled heating, the precursors were de-waxed at 800 °C for 2 hs in the air.

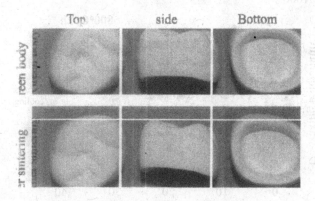

Fig. 2   Dental crown models of green bodies composed of photo sensitive acrylic resin with alumina particles formed by using the stereolithography and powder sintering processes.

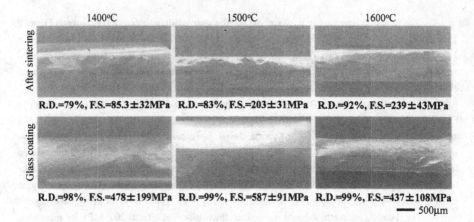

Fig. 3 Cross sectional ceramic microstructures of alumina flexural test samples with or without glass infiltrations. The plate specimens were formed by the stereolithography and heat treatment processes. RD and FS mean the relative density and flexural strength, respectively.

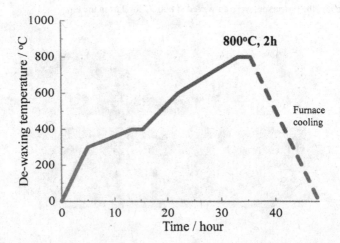

Fig. 4 A heat treatment pattern in air atmosphere to de-wax mixed photo sensitive acrylic resins in the alumina composite precursors fabricated by using the stereolithography

The measured densities of sintered alumina bodies with or without the glass coating (infiltration) are shown in Figure 5. While the densities without the coating were 2.61, 2.77, 3.02 and 3.29 g/cm^3 at 1400, 1500, 1600 and 1700 °C, respectively, those with the coating were 3.62, 3.67, 3.69 and 3.70 g/ cm^3, respectively. The flexure strengths for the alumina specimens were shown in Figure 6. The average flexural strengths without coating were 41.7, 94.6, 174 and 252 MPa at 1400, 1500, 1600 and 1700 °C, respectively, while those with the coating were 404, 640, 600 and 670 MPa, respectively. Through the grass infiltration, the sintered alumina can exhibit higher mechanical properties. The maximum flexural strength is about 670 MPa.

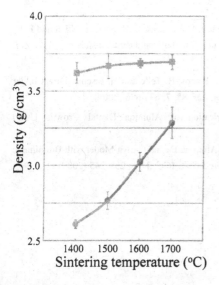

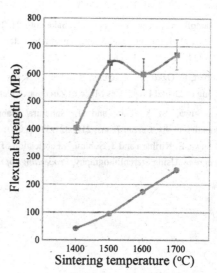

Fig. 5　The densities of alumina bodies with or without glass infiltrations.
●: Sintered body, ■: Glass coated ones

Fig. 6　The flexural strengths of the alumina bodies with or without the glass infiltrations.
●: Sintered body, ■: Glass coated ones

CONCLUSION

Alumina dental crown modes were fabricated successfully fabricated by using laser scanning stereolithography. Photosensitive acrylic resin composites and de-waxing heat treatment patterns were optimized to prevent macro crack formations. Through infiltration processes of grass materials into sintered alumina bodies, maximum flexural strength was obtained at about 670 MPa as an acceptable level for dental crown using.

ACKNOWLEDGMENTS

This study was supported by Priority Assistance for the Formation of Worldwide Renowned Centers of Research - The Global COE Program (Project: Center of Excellence for Advanced Structural and Functional Materials Design) from the Ministry of Education, Culture, Sports, Science and Technology (MEXT), Japan.

REFERENCES

[1] M. L. Griffith and J. W. Halloran, Freeform Fabrication of Ceramics via Stereolithography, *J. Am. Ceram. Soc*, **79**, 2601-2608 (1996)

[2] J. Stampfl, H. C. Liu, S. w. Nam, K. Sakamoto, H. Tsuru, S. Kang, A. G. Cooper, A. Nikel and F. B. Prinz, Rapid prototyping and Manufacturing by gelcasting of metallic and ceramic slurries, *Mater. Sci. Eng*, **A334**, 187-192 (2002)

[3] J. Ebert, E. Ozkol, A. Zeichner, K Uibel, O. Weiss, U.Koops, R. Telle and H. Fischer, Direct Inkjet Printong of Dental Prostheses Made of Zirconia, *J. Dent. Res*, **88**, 7, 673-676 (2009)

[4] M. Suwa, S. Kirihara and T. Sohmura, Fabrication of Alumina Dental Crowns Using Stereolithography, *Ceram.Trans*, **219**, 331-336 (2009)

[5] M. Suwa, S. Kirihara and T. Sohmura, Fabrication of Alumina Dental Crown Model with Biomimetic Structure by Using Stereolithography, *Proceedings of the 34th ICACC*, **31**, 8, 239-245 (2009)

LARGE-SIZED STRUCTURAL CERAMIC MANUFACTURING BY THE SHAPING OF THIXOTROPIC SLURRIES

Eugene Medvedovski
Umicore Thin Film Products
Providence, RI, USA

ABSTRACT

Manufacturing of large-sized structural ceramics is quite challenging, especially, if the ceramic components are designated for wear-, corrosion- and thermal shock-resistance applications. In many cases, monolithic complicated shapes, such as pipes, cones, cyclones, elbows and other components with sufficient wall thickness (up to 50-75 mm) are required in industry. The shaping from thixotropic slurries contained particles of various sizes allows processing complicated large-sized (up to 1 m diameter and height) components for structural applications. This technology is versatile, and it does not require expensive and complicated equipment. As example, silicon carbide-based ceramics obtained by the proposed technology are described.

INTRODUCTION

Large monolithic ceramic and refractory components with a high level of physical properties, mainly with mechanical and thermal properties, are highly required in different industries where processes under extreme conditions are occurred. In particular, large size monolithic ceramic components are required for mining and mineral processing, metallurgical, chemical and petrochemical industries and power generation. Significant losses due to wear-, corrosion- and thermal problems of equipment in mining and mineral processing, e.g. in coal or oil processing (oil with sand or ores), when the materials have to be transported through different piping systems, are estimated as more than $5 billion annually (NRC Canada estimation). Mineral extraction processes often involve the use of chemical treatments in strong corrosive environments at elevated temperatures and pressures, which create strong impacts on the equipment. These noted losses include damage, maintenance and repair of processing, classification, extraction equipment and related unscheduled shutdowns due to the actions of hard particles, high flow velocities with changeable flows and turbulence (i.e. associated with abrasive and erosive destruction of the equipment), necessity to use corrosive media, high temperature processes and occurred thermal shocks with often situations when combinations of these harsh conditions take place. Similar problems are often occurred in processing of cements, fertilizers and in chemical industry. Metallurgical operations include the necessity to use and maintain casting ladles and crucibles, mixing devices, lining of furnace, combustion chambers, converters and some other high-temperature heating units, which are also undergone by corrosion (e.g. molten metals and slags) and wear combined with high temperatures and thermal shocks. Coal-fired power generation plants, gasifiers and incinerators have similar harsh environments, when refractory lining and processing equipment are subjected by the severe destruction. It is obvious that "traditionally" used metals and alloys, as well as polymers, cannot withstand these conditions, and, therefore, the components from appropriate ceramic materials have to be manufactured.

Large size ceramic and refractory components, which often have to be with complicated shapes, e.g. pipes, cones, reducers, elbows, Y- and T-sections, crucibles, special shape bricks and many others, have to be monolithic with dimensions up to 1 m in diameter and height with minimal joints, and manufacturing of these components is really challenging. This manufacturing also has to be rather inexpensive and versatile due to the needs of variety of shapes. Because of the necessity of large and complex shapes of ceramics, minimal or even zero shrinkage has to be maintained during processing steps (e.g. in firing) that will promote the minimization or absence of cracking and other defects occurrence.

In order to produce these large and complicated ceramic components, several processing methods can be considered. These methods include uniaxial and cold isostatic pressing, extrusion, slip and gel casting and some others. However, all these methods have serious limitations dealt with a possibility to produce either only simple shapes (e.g. uniaxial pressing and extrusion) or necessity to use very expensive and high maintenance equipment (cold isostatic pressing) and necessity to use very high pressures (pressing and extrusion), etc. Traditional fine-grain oxide, carbide and nitride ceramics for these applications may be produced only with rather simple shapes. Shrinkage minimization is a real challenge for all "traditional" methods and fine-grain materials. Hence, a formation of ceramic structures with specially selected or formulated grain size compositions providing high compaction, zero or almost zero shrinkage and high mechanical and thermal properties, as well as a special forming method that has to be versatile, i.e. with a possibility to use a variety of molds, have to be utilized for the noted purposes.

In order to overcome the mentioned processing issues, in particularly, related to formulations and processing of "traditional" oxide and non-oxide ceramics, thixotropic casting technology can be applied. This technology can encompass appropriate compositions of ceramics and casting of the slurries based on these compositions utilizing their thixotropic behavior, i.e. creating a flow behavior to certain viscous slurries by mechanical activation (agitation) and a solidification of these slurries when the mechanical action is over. In this case, inexpensive molds providing complex shapes can be fabricated rather quickly specifically for each particular custom design.

This paper summarizes the development and manufacturing results obtained working with different ceramic and refractory materials and large-size products during a number of years. It demonstrated the applicability of the thixotropic casting technology for the appropriate ceramics, which were successfully served in industry for wear-, corrosion- and thermal shock-protection and high temperature applications.

PRINCIPLES OF CERAMIC COMPOSITIONS AND THIXOTROPIC CASTING TECHNOLOGY

Some basic principles of the technology using thixotropic casting, particularly for silica-based and some other oxide ceramics, were proposed by Yu. Pivinsky[1-3] and applied by him and some other researchers[1-8]. These principles, compositions and technology were further applied and modified for particular ceramics also by the author of this article. Basically, the technology may be applied for different ceramic materials - oxides, non-oxides and their combination, such as alumina, mullite, zircon, spinel, silica, silicon carbide, silicon nitride and others. In particular, the ingredients with high hardness, chemical inertness and refractoriness are selected for manufacturing of the components for wear-, corrosion- and thermal shock resistance applications. Ceramic compositions, which may be suitable for thixotropic casting technology of large-sized and complex-shape components, have to contain several ingredients, including coarse-sized powders (with particles larger than 0.5 mm), relatively fine powders (micron-size level) and ceramic binding system. The coarse ingredient may be of one, two or even more powders with different particle sizes, which are selected from the general dimensions of a ceramic product standpoint. The combination of the particles of different sizes, including coarse and fine powders, taken in the selected proportions will provide a high level of particle compaction. In general, coarse ingredients create a "skeleton" of the ceramic composite system, while fine-grained ingredients not only fill the space between larger particles but also reinforce the system through the "bridging" effect.

Because of the considered materials are designated for the applications where thermal, corrosive and wear actions and their combinations may take place, ceramic formulations have to satisfy special requirements. The principal grains of the mineral compositions have to have high hardness or refractoriness or inertness to strong corrosive environments (or combination of these properties). A high level of the grain compaction may be achieved, as mentioned, by the optimization of particle size distribution (it also may be additionally improved due to the features of manufacturing process). The bonding phase has to have high fracture toughness and strength. The principle grains and the ingredients for the ceramic bonding phase have to be taken in the ratios providing not only uniform distribution of the bonding phase, but the bonding phase also has to promote interaction between the composite ingredients during firing.

A special role in the ceramic compositions for thixotropic casting is "devoted" to the ceramic binding phase. This ceramic binding phase, that fills the space between the formed "skeleton" of the coarser and finer ingredients, also provides the specific casting ability, and it is responsible in the greater extent for the hardening of the ceramic body during high-temperature firing, and it actually promotes interaction between the major "skeleton" ingredients. This binding ingredient is a specially prepared colloidal system (ceramic slurry) consisting of solid and liquid phases, which altogether promote the creation of unique properties suitable for thixotropic casting.

This binding system that may be named as highly concentrated ceramic bonding systems HCCBS, in order to be effective for manufacturing of ceramic bodies, in particular, with large complex shapes for severe application conditions mentioned above, should satisfy the strict requirements. They should provide high adhesion properties to the ceramic fillers and should provide significant strength in hardening. During ceramic processing, the slurries based on HCCBS should prevent settling of coarse ingredients (fillers) of the ceramic composition, i.e. they have to have high sedimentation stability. Low or, preferentially, zero shrinkage during firing is very important, especially for large complex shape bodies manufacturing. No softening and decomposition should occur at heating and high-temperature cycling. HCCBS after firing should be chemically inert and should provide required mechanical and thermo-mechanical properties of the ceramics. Usually, the content of HCCBS in the ceramic compositions varies from 10 to 50 vol.-% depending on the type of ceramics, required manufacturing and application properties and particle sizes of the coarser ingredients.

HCCBS for thixotropic casting have positive points and some limitations with the features of hardening principles and mechanisms. They may be classified as follows:

Clay-based systems provide hardening during firing. They are inexpensive, and promote casting and sintering of ceramics. However, these materials do not provide high mechanical strength at the green state if the contents of the clay-based binding system is not sufficient and hardening is rather slow. Also, they may limit a working temperature and be responsible an excessive deformation at elevated temperatures. They also promote sufficient amounts of a glassy phase formation during firing that may be not very favourable for the service in highly corrosive and abrasive environments, and elevated firing shrinkage.

Binding systems based on alkali silicates (e.g. liquid glass) or ethyl silicate or colloidal silica harden due to chemical reaction at elevated temperatures. Strengthening in these systems occurs due to cross-linking by polycondensation mechanism and transformation of -Si-OH bonds into siloxane -Si-O-Si- bonds. These materials have similar benefits and limitations as the clay-based materials.

High-temperature cements (e.g. calcium aluminates, high-alumina cements and some others) harden at room temperatures and promote additional strength at elevated temperatures. Also sub-micron and, especially, nano-sized alumina powders may be successfully used. The compositions with cements demonstrate limited deformation at high temperatures, especially with high-alumina cements, minimal shrinkage and high thermo-mechanical properties. However, high-temperature interaction between this binding system and the major phases is limited, i.e. micro-cracking is possible during manufacturing.

Some salts (e.g. chlorides) of the elements providing a formation of high-refractory oxides, e.g. aluminum chloride, magnesium chloride and some others, also demonstrate binding effect, and, in the appropriate combination with the major phases, provide acceptable thermo-mechanical properties. However, the hardening rate and green strength for the compositions based on these binding systems are not very high. Besides, volatilization of chlorine ions is not favourable from the health and safety standpoint.

Phosphate-based binding systems, e.g. aluminophosphate, chromphosphate, aluminochromphosphate, magnesiaphosphate and others, harden at room and elevated temperatures, and they provide additional strength at elevated temperatures. The strengthening in these systems occurs due to association of phosphate molecules by strong hydrogen bonds and polymerization through P-O bonds resulting in the interaction of phosphate ions, e.g. $(PO_4)^{3-}$, with refractory fillers. These systems demonstrate a high level of green strength, minimal shrinkage during manufacturing, limited deformation at elevated temperatures. Good adhesion of phosphates to the high-temperature fillers promotes interaction between the ingredients and, as a result, high thermo-mechanical properties and corrosion resistance.

A special group of the HCCBS may be selected and prepared specifically for the different major phases, e.g. oxides, non-oxides or their combinations. In particular, the slurries are based on nano- and sub-micron or a mix of micron and sub-micron ceramic powders dispersed in a liquid phase with special properties.

Preparation of HCCBS, e.g. for the last group, includes preparation of the slurry contained dispersed ceramic powders of micron and sub-micron sizes in a liquid phase (usually water with addition of some agents providing powder dispersion, pH regulation and binding properties). These slurries are prepared by the milling of the ceramic ingredients in the liquid media with consequent slurry adjustment if it is required.

The prepared colloidal slurry, particularly for large size bodies manufacturing, is mixed with coarser ingredients to obtain working suspensions with thixotropic properties. In some cases, working thixotropic suspensions may be prepared by the mixing and milling of all ingredients (sub-micron, micron and coarse particles). The thixotropic slurry is cast into plaster or, in some cases, metallic or plastic molds. The cast bodies are demolded, dried and then fired at temperatures providing required physical properties. A general schematic of the process is shown in Fig. 1. This technology is rather simple and versatile, it does not require expensive and high-maintenance equipment, and it is well suitable for manufacturing of custom-design ceramic components.

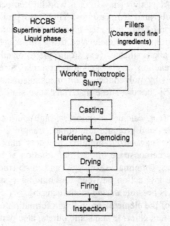

Fig. 1. General schematic of the thixotropic casting process

During preparation of HCCBS and the casting slurries, some important technological "tools" may be used in order to promote manufacturing. For example, mechanical activation of the solid ingredients of HCCBS may be applied. This action changes free energy of the system by mechanical force and reduces the particle sizes of the fine colloidal ingredients; it also promotes amorphization of the surface of some ingredients and increases adhesion and interaction of the solid-liquid phase. In some cases, mechano-chemical activation may be applied that promotes the surface interaction between the solid and liquid ingredients. Mechano-chemical activation is also very effective in the case of the use of phosphate-based binding systems. HCCBS have to be stable during storage and handling, and they have to possess required physical properties, e.g. certain viscosity at high solid content in order to prevent settling of the solid particles, which may be of large sizes (up to 1-2 mm), and to promote thixotropic properties of the working slurry.

The preparing working slurries usually have rather low liquid phase contents (only 7-15 wt.-%; in some cases, even lower). One of the key factors related to processing of large ceramic components using HCCBS is the thixotropic behaviour of the prepared slurries. Casting of the ceramic components may be conducted in a "self-flowing" condition. However, in order to "activate" the high-solid content slurries, i.e. in order to promote their flowability, casting may be conducted under vibration. The influence of mechanical activation, e.g. vibration, on slip viscosity may be seen on Fig 2. In general, the dependence of viscosity (η) of concentrated slurries vs. solid content is in accordance to the Krieger-Dougherty model:

$$\eta = \eta_{lp} \left(1 - \Phi/\Phi_m\right)^{-n},$$

where η_{lp} is viscosity of the liquid phase, Φ is a solid volume fraction, Φ_m is max. packing of solid fraction in the slurry, respectively, and n is an empirical coefficient usually taken as 2-2.5 for spherical particles. It can be seen from Fig. 2 that, in the case of mechanical activation of thixotropic slurries, viscosity does not grow very fast with increase of solid fraction. The remarkable flowability of these slurries provides the filling of the large and complex-shape molds. The applied vibration also promotes better compaction of the particles with different sizes resulting in rather dense and "homogeneous" (i.e., in this case, without delamination) bodies. Rather quick increase of viscosity and hardening when the activation is stopped with consequent increase of the strength of the cast bodies are provided by the thixotropic thickening.

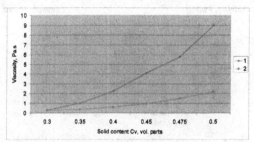

Fig 2. Viscosity of the slurries based on high-alumina sub-micron powders with some additives-"modificators"
1- thixotropic thickening; 2- after mechanical activation

Particle size distribution in thixotropic slurries is an important factor for manufacturing of large-size and complex-shape ceramic bodies, and it has to be specially selected in order to attain required processing and physical properties of the ceramics. The thixotropic slurries are distinguished by rather lower contents of the "superfine" (colloidal) ingredient. Opposite to the "traditional" approach of manufacturing of refractory and thermal shock resistant compositions, where the content of the "superfine" ingredient is 60-80%, the content of this ingredient for thixotropic casting is only 15-35%; however, the desirable particle size of this ingredient is below than 10 μm, and the presence of sub-micron particles (in some cases, also nano-particles) is highly important. Typical particle size distributions of the "traditional" and thixotropic cast bodies are showed in Table 1.

Table 1. Particle Size Distribution for Ceramic Manufacturing Based on Different Slurries

Particle size	"Traditional" approach	Thixotropic slurries
<0.06 mm	60-80%	15-35%
0.06-0.5 mm	15-25%	10-20%
>0.5 mm	3-15%	50-65%
(0.5-3 mm)		
	good for small-mid. size bodies	good for mid.-large size bodies

Basically, particle size distribution of the ceramic composition (granulometric composition) is selected based on the necessity to attain a high level of compaction, ability to prepare the stable slurries without particles segregation with good flow, and these slurries should fill cavities of large complex-shape molds. As mentioned above, granulometric compositions and particle sizes are also selected according to dimensions and shapes of the producing components, i.e. large bodies require the presence of particles with larger sizes. According to these requirements, multi-fractional granulometric compositions are preferred. They provide higher mechanical properties in comparison with the ceramic compositions based on the particles with narrow particle size distribution due to better compaction. As an example, the comparison of density and strength of the high-alumina ceramics based on narrow particle size distributions and multi-fractional compositions is shown in Fig 3. All the samples (bars with dimensions of 10x10x100 mm) were fired at 1650°C. It can be seen that the samples based on specially selected particle size distribution had higher density and, especially, strength comparing them with the samples based on mono-fractional compositions.

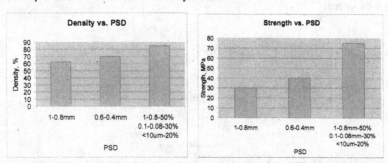

Fig 3. Density and flexural strength of high-alumina ceramics based on different granulometric compositions

Excessive amounts of the "fine" ingredients in ceramic compositions may result in local sintering and elevated shrinkage during firing that, consequently, in weakening of the bonds in the skeleton and finally in decrease of strength and higher deformation of ceramics at elevated temperatures. The influence of the content of the bonding (HCCBS) phase in the "coarse" ceramic composition on some technological and mechanical properties of alumina ceramics with mullite bonding phase is illustrated in Table 2.

Table 2. Properties of Alumina-Mullite Ceramics with Different Contents of the HCCBS Phase

Bonding phase (HCCBS), %	Shrinkage, %	Density, g/cc	Strength, MPa	Temperature of deformation, °C
15	0.5	3.2	35	1640
30	1.2	3.1	28	1600
45	3.4 (cracking is observed)	3.05	18	1520

SILICON CARBIDE-BASED CERAMICS MANUFACTRURED BY THIXOTROPIC CASTING

Manufacturing of SiC-based ceramics is a good example demonstrating the applicability of the thixotropic casting technology that allows to attain high physical properties of ceramics, which are well suitable for severe application conditions. The ceramics have been developed in the systems, such as SiC, SiC-Al$_2$O$_3$, SiC-Si$_3$N$_4$, SiC-Si$_3$N$_4$-Al$_2$O$_3$, SiC-Al$_2$O$_3$-SiO$_2$ and some others; some ingredients in rather small quantities may be added to improve densification and properties of the ceramics. The major ingredient (SiC) was represented by the particles with different grain sizes with selected multi-fractional granulometric compositions; the SiC particles were ranged from 3-10 μm to 1-2.5 mm. The selected granulometric compositions provided rather good compaction due to smaller particles filled the spaces between larger particles formed the "skeleton", as well as prevented the segregation of the particles with different sizes.

As the binding ingredient, alumina, alumina-mullite or alumina-silicon carbide mixtures were used. These mixtures based on micron-submicron starting powders were prepared in accordance with the above mentioned principles using wet (colloidal) processing resulting in the HCCBS formation. Some of the compositions contained small amounts of specially selected sintering aids, which promoted densification of the ceramics. In the case of manufacturing of "pure" SiC ceramic components, the binding phase also contained SiC powders of the micron sizes. The prepared slurry with a high solid content (>85 wt.-%) and with an "appropriate" level of viscosity was mixed with coarser ingredients in accordance with the mentioned principles. The prepared thixotropic slurries were cast into plaster molds, solidified, and the obtained bodies after demolding were dried and fired at the temperatures in the range of 1500-1550°C. The ceramic formation occurred in accordance with "reaction-bonding" principles based on the partial oxidation of the surface of the particles of the major ingredients and interaction of the occurred new phases with some other ingredients of the ceramic composition, which were earlier described[9, 10].

The obtained SiC-based ceramics had heterogeneous structures, which consisted of SiC (in some cases, also Si$_3$N$_4$) grains of different sizes bonded mostly by the aluminosilicate microcrystalline (e.g. mullite)-glassy phase that is rather uniformly distributed between larger grains. The examples of microstructures of the SiC-based ceramics with different starting SiC ingredients and with different bonding phases (e.g. SiC-Al$_2$O$_3$-SiO$_2$ and some others) are shown in Fig 4. When Si$_3$N$_4$ was used as one of the ingredients in ceramic compositions, the bonding phase also contained SiON and SiAlON. The formed new phases due to *in-situ* reaction bonding fill the pores between the major SiC grains. Due to the selected granulometric compositions and the reaction-bonding mechanism of the ceramic structure formation, the obtained ceramics had zero shrinkage that is favorable for manufacturing of large-size complex shape components.

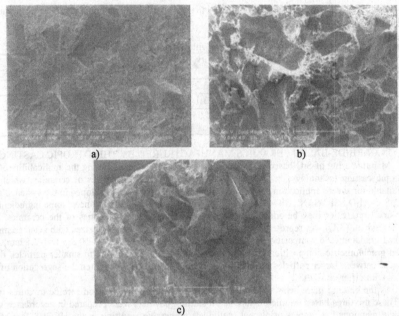

Fig 4. Microstructure of some SiC-based ceramics with different granulometric compositions
   a) Ceramics SiC-Si$_3$N$_4$-Al$_2$O$_3$ (magnification 500x)
   b) Ceramics SiC-Al$_2$O$_3$ (magnification 60x)
   c) Ceramics SiC (magnification 35x)

   Due to the features of the ceramic composition and firing, the surface of ceramic components is denser that is dealt with a higher level of the formation of the silicate glassy phase (almost zero open porosity), while the central area of the ceramics had some porosity. Small amounts of the additives used in some ceramic compositions promoted densification not only on the surface but also in the central area that consequently resulted in the physical properties increase. The surface densification, smoothening and strengthening are achieved by the optimization of the firing temperature and firing conditions (Fig. 5). The formed crystalline-glassy phase with elevated amounts of the glass on the surface also promotes the healing of the microcracks, which may occur due to the mismatch of coefficients of thermal expansion of the ceramic ingredients during firing. In the case of addition of alumina or alumina-mullite ingredients to ceramic compositions, densification and properties of the ceramic surface were increased (Fig. 5).

   The obtained SiC-based ceramics have a high level of physical properties (Table 3), which are well suitable for the wear-, corrosion- and thermal shock-resistant applications. Properties of the ceramics are defined by mineral composition (nature and content of major ingredients and additives), granulometric composition (e.g. presence, content and size of the largest particles, particle size distribution), bonding phase composition and content, processing features, etc. For example, the compositions with SiC of the largest particle size of 1-2 mm, as expected, have lower mechanical strength than the compositions with SiC of the largest particle size of 0.5-0.8 mm (denoted as

ABSC30m); however, the latter can be used only for the manufacturing of rather small or medium size components with thin walls. The compositions contained Si_3N_4 (with small-medium particle size) and without presence of coarse SiC particles (denoted as ASN) provided relatively high mechanical properties.

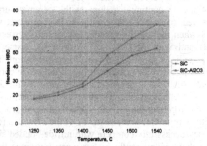

Fig 5. Influence of firing temperature and composition of the bonding phase on surface hardness of SiC-based ceramics

All the compositions possess excellent thermal shock resistance; the ceramic samples (bars with dimensions of 20x20x100-150 mm or tiles with dimensions of 100x100x20 mm) withstand repeatable thermal shocks of several hundred degrees – $10°C$ (hot air-water) without cracks. High thermal shock resistance of these ceramics is confirmed by the long-term service in actual severe application conditions. High thermal shock resistance of these ceramics is defined by the selected granulometric compositions and related microstructures, by the presence of the $SiC-Al_2O_3-SiO_2$ or $SiC-Si_3N_4-Al_2O_3-SiO_2$ crystalline-glassy bonding phase and by rather high thermal conductivity. The deviations from the appropriate ratios of the starting ingredients, e.g. in HCCBS formulation and content, or from the thixotropic casting parameters or from the firing conditions result in a lack of or an excessive formation of the bonding phase and phase segregation and, consequently, micro-cracking formation and properties decrease.

Table 3. Physical Properties of Selected SiC-based Compositions

Properties	ABSC20	ABSC15	ABSC17	ABSC30m	AS	ASN
Density, g/cm^3	2.85-3.0	2.85-2.95	2.85-2.95	3.1-3.2	3.0-3.2	2.7-3.0
Flexural strength, MPa	25-28	23-26	23-26	35-45	100-120	130-155
Rockwell hardness HRA	50-60	50-60	50-60	50-60	58-65	65-77
Thermal conductivity, W/m.K	23-26	23-26	23-26	25-30	30-35	35-40

The obtained SiC-based ceramics withstand the actions of corrosive environments, such as acidic conditions, high-temperature corrosive gases, including sulfur- and carbon-containing substances, and some others used in some chemical industrial applications, in the conditions of the oil refinery processing and combustion environment. The materials also withstand the actions of molten metals and alloys (e.g. Al, Zn) and slags, e.g. in the slagging gasification environment where corrosion, high temperatures and thermal shocks take place[11]. The ceramics contained Si_3N_4 along with SiC possess higher corrosion resistance, particularly in molten materials environment. Due to the presence of oxide-based bonding phase, the ceramics serve well in high-temperature (greater than $1400°C$) highly oxidizing environment where more "traditional" non-oxide ceramics experience difficulties related to high-temperature oxidation. During the service of these oxide-bonded SiC-based ceramics contained the oxide protective layer in high-temperature oxidizing environment, only insufficient further oxidation of SiC (and Si_3N_4 in the case of its presence in the ceramic composition) occurs but without fast structural transformation and mechanical degradation.

The obtained ceramics demonstrated excellent wear resistance in sliding abrasion conditions when the abrasive particles have a continual action on the ceramic surface either in dry or wet conditions. The results of the wear resistance tests conducted in accordance with ASTM G65 (action of silica sand in dry conditions when the abrasive media is supplied between a ceramic sample and a rotating rubber-lined wheel) and with ASTM B611 (action of coarse alumina particles in wet conditions when the abrasive media is supplied between a ceramic sample and a rotating steel wheel) for some SiC-based ceramics designated for manufacturing of large-size components are shown on Fig. 6. High wear resistance (small wear loss under the action of abrasive media) of the ceramics is dealt with optimized composition and technology, e.g. a high level of compaction of hard SiC grains bonded by the crystalline-glassy phase, firing conditions, etc. Modification of granulometric composition and small amounts of some additives promote wear resistance of the ceramics (e.g. ceramics ABSC15 and ABSC17). These ceramics can successfully compete with nitride-bonded silicon carbide (NBSC) and some alumina ceramics; they have significantly higher wear resistance than basalt widely used for large size components for wear-resistance application and, moreover, than steels (steel samples had very fast destruction at the testing in wet conditions, such as ASTM B611). Also these ceramics demonstrated high erosion resistance, e.g. withstanding the action of hard SiO_2 particles at different impingement angles. The mechanism of the wear process and more detailed data of wear resistance test results were described earlier[9, 12].

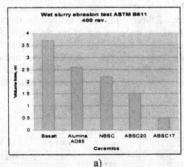 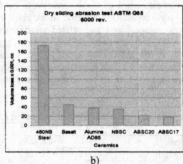

a) b)

Fig. 6. Wear resistance test results for some SiC-based ceramics – the smaller the volume loss the higher the wear resistance
   a) Wet slurry abrasion test ASTM B611 (400 revolutions)
   b) Dry sand rubber-lined wheel abrasion test ASTM G65, procedure A (6000 revolutions)

The developed compositions and technology allowed to manufacture different products with complex shapes, e.g. pipes, cones, cyclones, vortex, elbows and others (see Fig. 7). The ceramic compositions contained large grains of 1-2.5 mm with optimized particle size distribution (e.g. ABSC20, ABSC15 and ABSC17) were successfully used for large size and complex shape monolithic components with diameter or/and height up to 1 m, and wall thickness 20-50 mm and even greater. Manufactured ceramics were successfully used as the components for the service in severe wear-, corrosion- and thermal shock conditions, e.g. in mining and mineral processing, oil extraction processing (when processing oil contained large amounts of abrasive sand and some other minerals), coal-fired and gasification processing, in incinerators, metallurgical processing (in contact with molten materials), as kiln furniture components (e.g. supporting bricks) and many other industrial situations. Due to the features of the composition and technology, these oxide bonded SiC-based ceramics with well-acceptable performance is significantly less expensive than "conventional" nitride- or reaction-bonded SiC ceramics and, moreover, than hot-pressed or pressureless sintered SiC ceramics, which also have serious limitations in processing of large size products.

Fig. 7. Different SiC-based ceramic products manufactured through thixotropic casting

CONCLUSIONS

Thixotropic casting using inorganic highly-concentrated ceramic bonding slurries was successfully applied for commercial manufacturing of numerous structural ceramic components with a high level of wear-, thermal shock- and corrosion resistance. The technology is efficient and versatile, inexpensive, and it provides manufacturing of large-sized and complex shape components, which were successfully worked in industrial applications. Adequate properties of ceramics are achieved due to the combination of selected granulometric composition, shaping and compaction through thixotropic casting and high-temperature bond formation between major crystalline phases of ceramics.

REFERENCES

[1] Yu.E. Pivinsky, Fundamentals of Technology of Ceramoconcretes, *Refractories*, N.2, 34-42 (1978) (in Russian)

[2] Yu.E. Pivinsky, Refractory Materials and Concretes Based on Ceramic Binding Suspensions, *Refractories*, N.6, 54-55 (1981) (in Russian)

[3] Yu.E. Pivinsky, Ceramic Binders and Ceramic Concretes, Moscow, 1990 (in Russian)

[4] P.L. Mityakin, Yu.E. Pivinsky, Properties of Quartz Ceramic Castables, *Refractories*, N.9, 55-59 (1980) (in Russian)

[5] Yu.E. Pivinsky, V.A. Bevz, Fabrication of Aqueous Zircon Suspensions and the Study of their Rheological, Technological and Binding Properties, *Refractories*, N.8, 38-43 (1979) (in Russian)

[6] Yu.E. Pivinsky, V. A. Bevz, Preparation of Aqueous Suspensions of Mullite and the Study of their Rheological and Technological Properties, *Refractories*, N.3, 45-50 (1980) (in Russian)

[7] S. Chaudhuri, Monolithic Ladle Linings, *Interceram*, **43** (6) 478-480 (1994)

[8] S. Yuan, Self-Flowing Castables with Ultra-Low Cement Content, *Interceram*, **45** (4) 244, 246, 248 (1996)

[9] E. Medvedovski, Silicon Carbide-Based Wear-Resistant Ceramics, *Interceram*, **50** (2) 104-108 (2001)

[10] E. Medvedovski, On High Temperature Oxidation of Some Carbide and Nitride Ceramic Bodies, *Interceram*, **56** (3) 168-173 (2007)

[11] E. Medvedovski, R.E. Chinn, Corrosion Resistant Refractory Ceramics for Slagging Gasifier Environment; p. 547-552 in *Ceramic Engineering and Science Proceedings, 2004*. Proceedings of The 28th Annual International Cocoa Beach Conference on Advanced Ceramics & Composites, Cocoa Beach, FL, January 25-30, 2004

[12] E. Medvedovski, Wear-Resistant Engineering Ceramics, *Wear*, **249**, 821-828 (2001)

ORIENTED ALUMINA CERAMICS PREPARED FROM COLLOIDAL PROCESSING IN MAGNETIC FIELD

Satoshi Tanaka[1] Atsushi Makiya[1] and Keizo Uematsu[1]

[1]Dept. Materials Science and Technology, Nagaoka University of Technology
1603-1 Kamitomioka, Nagaoka Niigata, 9402188 JAPAN

ABSTRACT
Influence of particle size on orientation was examined in colloidal processing in magnetic field, and sintering effects of subsequent sintering was studied on the development of textured microstructures. Alumina particles with 0.5 and 1.3 μm were used. Dispersed slurry with 5-50 vol% in solid loading was prepared and formed in high magnetic field 10 T at room temperature. They were sintered at 1600°C. The orientation degree of the green body and sintered ceramics was evaluated by the polarized microscopy and XRD measurements. Microstructure in ceramics was observed by the SEM. Degree of orientation in powder compact was in the range of 40% - 80%. The degree of orientation of compacts made from particles 1.3 μm was higher than those made from small particles 0.5μm. On the other hands, the degree of orientation of the green compacts with smaller particles increased drastically by densification and grain growth during sintering processing.

1. INTRODUCTION
Texture consisting of crystallographically oriented grains drastically improves a variety of functional property in ceramics. A unique processing with high magnetic field has been developed to design it[1-3]. The processing involves colloidal process in the magnetic field and subsequent sintering, and utilizes the anisotropic magnetic susceptibilityof crystals in non-cubic systems[4,5]. The advantage of the method is that it is capable of aligning the crystalline axis even in conventional fine particles with near-spherical shape. They allow easy densification and microstructure development for oriented texture during the subsequent sintering. We have reported successful application of the method for several materials such as alumina[4-6], titania[7,8], bismuth titanate family[9,10], zinc oxide[11] and tungsten bronze systems[12-14]. Process of texture development should be understood.

The driving force for orientation is the minimization of magnetization energy in a magnetic field. The crystal axis of the highest diamagnetic susceptibility is oriented normal to the magnetic field. The magnetic torque under the driving force is given by the following equation.

$$T = \frac{1}{2\mu_0} \Delta\chi B^2 \left(\frac{4\pi r^3}{3} \right) \sin 2\theta \qquad (1)$$

Fig. 1 Raw powders (a) 1.3μm (AA1), (b)0.5μm (AA03)

159

where, $\Delta\chi$ is anisotropy of magnetic susceptibility, r particle radius and B magnetic flux density, which govern the magnetic torque T.

Besides, the past papers have been reported that oriented microstructure was also developed by densification and grain growth of oriented particles during subsequent sintering[6,7]. The oriented particles with large size in green body should grow with incorporating fine random particles during sintering.

The objectives of this study are to examine the influences of particle size and sintering processing on the degree of orientation and texture development in alumina ceramics, which are prepared by colloidal processing in magnetic field.

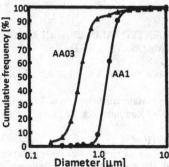

Fig. 2 Size distribution of each powder

2. EXPERIMENTAL

Green bodies with various particle orientations were prepared in colloidal processing in high magnetic field. Two commercial α-alumina powers (AA1 and AA03, Sumitomo Chemical Co., Japan) were used as raw materials (Fig.1). The alumina powders were nearly mono-dispersed and their shapes were spherical. Fig.2 shows the size distribution of each powder, which was measured by x-ray sedimentation method. Particle diameters were 1.3 and 0.5 μm. The powder was mixed with distilled water and a dispersant (ammonium polyacrylate; Seruna D305, Chukyoyushi Japan). The volume fraction of powder was 5–50 vol%. The mixture was ball milled for 24 h to make slurry and was poured 5ml into cylindrical container with 20mm in diameter. After dried in high magnetic fields 10T provided by a superconducting magnet, the compact was sintered at 1600°C for 1 hour.

The degree of particle orientation was evaluated on transparent powder compacts with a polarized light microscope[15]. We assumed that retardation of a compact $R_{compact}$ could be written as follows:

$$R_{compact} = d\frac{\rho}{100}\Delta n_{compact} \tag{2}$$

where $\Delta n_{compact}$ is the birefringence of compact, d the measured thickness of a thinned compact and ρ is the relative density; the product $d\rho/100$ corresponds to the net length of the solid through which the light travels. The measured relative density ρ was about 60% for all compacts. The degree of particle orientation f is defined as the ratio of birefringence of compact $\Delta n_{compact}$ and that of single crystal $\Delta n_{single\ crystal}$, as follows:

$$f = \frac{\Delta n_{compact}}{\Delta n_{crystal}}100 = \frac{\Delta n_{compact}}{\varepsilon - \omega}100 = \frac{\Delta n_{compact}}{0.0075}100 \tag{3}$$

where, the term $(\varepsilon-\omega)$ is the birefringence of single crystal along a-axis and is 0.0075 for alumina observed perpendicular to c-axis. The birefringence $\Delta n_{compact}$ was measured as follows.

A thinned sample (typically 0.3mm thick) was made transparent by addition of adequate immersion liquid (methylene iodide with sulfur, refractive index, RI = 1.77) having the same refractive index as alumina (RI = 1.76-1.77). The liquid eliminates the reflection of light at the liquid/alumina interface, and allows all light pass through the particle. A polarized light microscope was used to observe the optical anisotropy in a green compact under a Cross Nichols. A compensator of the Berek type was used for quantitative measurement of retardation.

Highly orientated alumina ceramics was examined by X-ray diffraction (XRD; MO3XHF, Bruker Japan) with Cu Kα radiation. The conventional 2θ/θ scan was used to obtain the X-ray diffraction profile. The Rocking curve analysis was measured to determine the distribution of preferred orientation. The degree of grain orientation f of sintered body was calculated by the calculated retardation which was analyzed by using probability distribution derived from the Rocking curves[16]. Detailed calculation methods had been reported ref.16.

$$f = \frac{\int_{-\pi/2}^{\pi/2} \int_0^{2\pi} \Delta n_{particle}(\theta,\phi) \cdot P(\theta) \cdot \sin\theta \cdot d\phi \cdot d\theta}{\int_{-\pi/2}^{\pi/2} 2\pi \cdot P(\theta) \cdot \sin\theta \cdot d\theta} \cdot \frac{1}{0.0075} \quad (4)$$

where, $\Delta n_{particle}$ is the theoretical birefringence of a grain, which was expressed as follows[16],

$$\Delta n_{particle} = \omega - \frac{a \cdot \varepsilon}{\sqrt{\omega^2 \cdot \sin^2\zeta + \varepsilon^2 \cdot \cos^2\zeta}}$$

$$= 1.7685 - \frac{1.7685 \cdot 1.761}{\sqrt{1.7685^2 \cdot \sin^2\zeta + 1.761^2 \cdot \cos^2\zeta}} \quad (5)$$

where, ζ is inclination angle between the polarized light and c-axis, which is given by $\zeta = \cos^{-1}(\sin\theta\sin)$. In eq.(4), $2\pi P(\theta)\sin\theta d\theta$ is probability of oriented particle at the small area in polar coordinate, which $P(\theta)$ was measured from the Rocking curve[16].

3. RESULTS and DISCUSSIONS

Fig.3 shows the degree of orientation in green body as a function of solid loading in slurry. The degree of orientation was almost the same 80% in the green bodies made from 5 and 30vol% slurries with 1.3μm (AA1) powder. It decreased at higher solid loading. The result indicates excellent dispersion of particles in slurries with low solid loading, and is consistent to the percolation theory. The degree of orientation was low for 0.5μm (AA03)

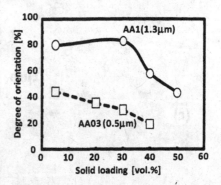

Fig. 3 Degree of orientation in green body as a function of solid loadings in slurry.

powder, and decreased with solid loading in the slurry. The magnetic torque, which is proportional to particle volume in eq.(1) is clearly responsible for this result.

Fig.4 shows the XRD patters of green and sintered bodies which were made from slurry 20vol% with 0.5μm (AA03) powder. XRD patterns were taken for the top surface of the specimens. Orientation increased a diffraction peak derived from *10$\underline{1}$0* in the green body prepared in the magnetic field. However, the diffraction peak *006* was hardly observed. After sintering at 1600°C for 1hr without magnetic field, diffraction peaks belonging to c-face *006* and *00$\underline{1}$2* were remarkably enhanced as shown in **Fig.4**(a). The direction and enhancement of orientation noted were the same with that of past studies[4-6].

Fig.5 shows SEM micrographs of microstructures in green and sintered bodies made from 1.3μm (AA1) and 0.5μm (AA03) powders. Green bodies were prepared from 20vol% slurry in the magnetic field 10T. The grain sizes increased markedly in sintered body prepared from 0.5μm (AA03) powder, whereas grain sizes increased only slightly for 1.3μm (AA1). **Fig.5** shows the development of texture in sintering. Past studies reported

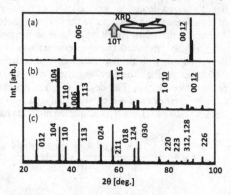

Fig. 4 XRD patterns of oriented alumina, (a) sintered at 1600°C for 1h, (b) green body and (c) random alumina green body made without magnetic field.

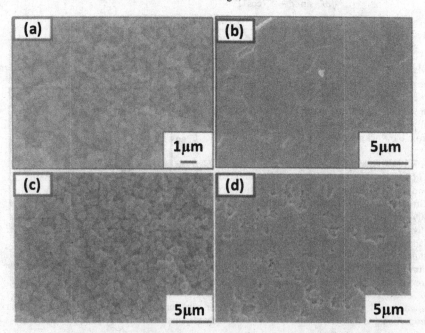

Fig. 5 Microstructure of green and sintered bodies, (a) green body made from 1.3μm (AA1) powder, (b) sintered body of (a), (c) green body made from 0.5μm (AA03) powder, (d) sintered body of (c).

that densification and grain growth of oriented grains contribute to the development of oriented structure in many systems[6-8,11].

Fig.6 shows the relationship between the degrees of orientation in green and sintered bodies. The solid loadings were controlled to prepare green bodies of various degrees of orientation as shown in **Fig.3**. With a fine powder, the degree of orientation increased and reached 80% after sintering even with low degree of orientation in starting state. In contrast, the degree of orientation remained almost constant after sintering for 1.3µm (AA1) powder. Drastic improvement of orientation is expected in grain growth as shown in **Fig.5**. Large particles, which are

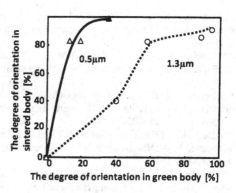

Fig. 6 Relationship between degree of orientation of green and sintered bodies.

characteristically well oriented, incorporate small particles of poor orientation in grain growth. Indeed, some large particles are noted in **Fig.2** for fine raw powder (AA03). A small amount of large particles clearly play a very important role in the texture development.

4. CONCLUSIONS

Particle size and sintering processing affected on the development of oriented microstructures, which was made in colloidal processing in magnetic field and subsequent sintering without magnetic field. The degree of orientation of compacts made from particles 1.3 µm was higher than those made from finer particles 0.5µm. On the other hands, the degree of orientation of the green compacts with smaller particles increased drastically by densification and grain growth during sintering processing.

REFERENCES

[1] J.G.Noudema, J.Beilleb, D.Bourgaulta, D.Chateignerc and R.Tourniera, Bulk textured BiPbSrC aCuO (2223) ceramics by solidification in a magnetic field, *Physica C*, 264, 325-30 (1996)

[2] Y.Nakagawa, H.Yamasaki, H.Obara and Y.Kimura, Superconductiong properties of grain-oriented samples of YBa$_2$Cu$_3$O$_y$, *J.J.Appl Phys.*, 28, 4, L547-50 (1989)

[3] M. H. Zimmerman, K. T. Faber, E. R. Fuller Jr., K. L. Kruger and K. J. Bowman, Texture assesment of magnetically processed iron titanate, *J.Am.Ceram.Soc.*, 79, 1389-93 (1996)

[4] K.Uematsu, T.Ishikawa, D.Shoji, T.Kimura and K.Kitazawa, Japan-Patent 3556886

[5] T.S.Suzuki, Y.Sakka and K.Kitazawa, Orientation amplification of alumina by colloidal filtration in a strong magnetic field and sintering, *Adv. Engineering Mater.*, 3, 44-46 (2001)

[6] A.Makiya, D.Shoji, S.Tanaka, N.Uchida, T.Kimura and K.Uematsu, Grain oriented microstructure made in high magnetic field, *Key Engineering Mater.*, 206-213, 445-48(2002).

[7] Y.Sakka and T.S.Suzuki, Textured development of feeble magnetic ceramics by colloidal processing under high magnetic field, *J.Ceram.Soc.Jpn.*, 113, 26-36 (2005)

[8] A.Makiya, K.Kusumi, S.Tanaka and K.Uematsu, Particle oriented titana ceramics prepared in a magnetic field, *J.Euro.Ceram.Soc.*, 27, 797-99 (2007)

[9]A.Makiya, D.Kusano, S.Tanaka, N.Uchida, K.Uematsu, T.Kimura, K.Kitazawa and Y.Doshida, Particle oriented bismuth titanate ceramics made in high magnetic field, *J.Ceram.Soc.Jpn*, **111**, 702-4 (2003)

[10]Y.Doshida, K.Tsuzuku, H.Kishi, A.Makiya, S.Tanaka, K.Uematsu and T.Kimura, Crystal-oriented $Bi_4Ti_3O_{12}$ ceramics fabricated by high-magnetic-field method, *J.J.Appl Phys.*, **43**, 6645-48 (2004)

[11]S.Tanaka, A.Makiya, Z.Kato, N.Uchida, T.Kimura and K.Uematsu, Fabrication of c-axis orientated polycrystalline ZnO using a rotating magnetic field and following sintering, *J.Material Res.*, **21**, 703-7 (2006)

[12]Y.Doshida, H.Kishi, A.Makiya, S.Tanaka, K.Uematsu and T.Kimura, Crystal-oriented La-substituted $Sr_2NaNb_5O_{15}$ ceramics fabricated using high-magnetic-field method, *J.J.Appl.Phys.*, **45**, 7460-64 (2006)

[13]S.Tanaka, A.Makiya, T.Okada, T.Kawase, Z.Kato and K.Uematsu, C-axis orientation of $KSr_2Nb_5O_{15}$ by using a rotating magnetic field, *J.Am.Ceram.Soc*, **90**, 3503-06 (2007)

[14]S.Tanaka, T.Takahashi, R.Furushima, A.Makiya and K.Uematsu, Fabrication of c-axis-oriented potassium strontium niobate ($KSr_2Nb_5O_{15}$) ceramics by a rotating magnetic field and electrical property, *J.Ceram.Soc.Jpn*, **118**, 722-725 (2010)

[15]S.Tanaka, A.Makiya, S.Watanabe, Z.Kato, N.Uchida and K.Uematsu, Particle orientation distribution in alumina compact body prepared with the slip casting method, *J.Ceram.Soc.Jpn*, **112**, 276-279 (2004)

[16]A.Makiya, S.Tanaka, D.Shoji, T.Ishikawa, N.Uchida and K.Uematsu, A quantitative evaluation method for particle orientation structure in alumina powder compact, *J.Euro.Ceram.Soc*, **27**, 3399-3406 (2007)

DENSIFICATION MECHANISMS OF YTTRIA-STABILIZED ZIRCONIA BASED AMORPHOUS POWDERS BY ELECTRIC CURRENT ASSISTED SINTERING PROCESS

Tatsuo Kumagai and Kazuhiro Hongo
Department of Mechanical Engineering, National Defense Academy
1-10-20 Hashirimizu, Yokosuka-shi
239-8686, JAPAN

ABSTRACT

Rapid densification behaviors (relative density; $\rho < 0.92$) of mechanically milled 3 mol% yttria-stabilized tetragonal zirconia polycrystals (3Y-TZP) and 3Y-TZP containing 20 mol% boron powders by electric current activated/assisted sintering have been investigated. Although the starting 3Y-TZP powder was amorphous, the crystallization of tetragonal zirconia solid solution (t-ZrO_{2ss}) was completed before the appearance of rapid densification, indicating the impracticability of densification via amorphous viscous flow. The n (stress exponent) values estimated by a hot-pressing kinetic equation increase with decreasing effective stress (i.e., increasing ρ), but n recalculated by taking into consideration the threshold stress (σ_0) show almost the constant value of 2. These recalculated values are quite similar to those of creep deformation with an accompanying σ_0 in the intermediate-stress region, suggesting that the rapid densification behavior is controlled by grain boundary sliding.

The addition of boron changes the densification behavior drastically. The estimated n values without considering σ_0 show $n=1.35$ in the low temperature region (T: 1120 K) and $n=2$ in the high temperature region (T>1120 K), suggesting that the boron-rich glassy phase coexisting with tetragonal (monoclinic at room-temperature) ZrO_{2ss} may relieve the stress concentration generated by grain boundary sliding of the matrix ZrO_{2ss} grains.

INTRODUCTION

Electrical current activated/assisted sintering (ECAS) [1] is one of the most effective methods for fabricating fully densified nanoscale yttria-stabilized tetragonal zirconia polycrystals (TZP) in an extremely short period of time. The obtained TZP bulks have been reported to show excellent strength and toughness at ambient temperature and workability at relatively low temperatures.[2] As opposed to many creep experiments[3,4] (i.e., plastic deformation without volume change), however, studies concerning the rapid densification mechanisms during ECAS have been limited. Recently, it was suggested that the rapid densification of the ZrO_2-3 mol% Y_2O_3 (3Y-TZP) powder during ECAS proceeds through grain boundary sliding as well as creep deformation, and the stress exponent (n) compensated for threshold stress and the apparent activation energy (Q) are $n=2$ and $Q=460\pm26$ kJ/mol, respectively.[5] However, as far as we know, the densification behaviors of 3Y-TZP containing a small amount of glassy phase, which is considered to be an effective additive for improvement of creep deformation, have not been reported.

In this study, we try to compare the rapid densification behaviors of 3Y-TZP and boron added 3Y-TZP powders, and discuss the densification mechanisms.

EXPERIMENTAL PROCEDURE

Two kinds of amorphous powders of TZP (ZrO_2-3 mol% Y_2O_3, High Purity Chemicals Co., Ltd, Saitama, Japan) and TZP-20 mol% B (amorphous B, Mitsuwa Chemicals Co., Ltd, Osaka, Japan) were prepared by ball milling system (P-6, Fritch, Idar-Oberstein, Germany). Hereafter the yttria content in zirconia is omitted for convenience and "TZP" stands for "ZrO_2 containing 3 mol% Y_2O_3"

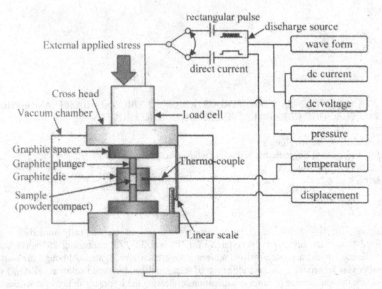

Fig. 1 Schematic diagram of the electrical current activated/assisted sintering system.

unless otherwise stated. The ball milled powders were consolidated by ECAS system[2] equipped with real-time monitors of current, voltage, pressure, temperature and cross head displacement (Fig. 1). Two gram of the powders were packed into a cylindrical graphite die (inner and outer diameter of 10 and 45 mm), and pressed by a set of graphite plungers (diameter of 10mm and height of 30mm) in vacuum chamber. During the consolidation of 999s, the applied stress and current are fixed to 46.5 MPa and a certain value of 550 to 900 A. The instantaneous shrinkage of the sample was obtained from the cross-head displacement, which is compensated for the thermal expansion of the graphite plungers obtained by blank test.[5] The instantaneous height of the samples was calculated by adding the instantaneous shrinkage (>0) to the final height measured by micrometer at ambient temperature. Naturally, the value of the final height at "high temperature" for each sample was compensated by using a linear thermal expansion coefficient of 3Y-TZP ($10.1 \times 10^{-6} K^{-1}$).[6] Assuming the weight and the diameter of the sample are almost constant during consolidation in the closed die, the instantaneous relative density was obtained by dividing the height of the fully densified sample (compensated for the linear thermal expansion) by the instantaneous height of the sample.

The microstructures were examined by X-ray diffractometry (XRD, M03XHF22, Mac Science Co., Ltd, Kanagawa, Japan) using CuKα radiation operated at 50 kV and 32 mA, field emission scanning electron microscopy (FE-SEM; S-4500, Hitachi Co., Ltd, Tokyo, Japan) and field emission transmission electron microscopy (FE-TEM; HF-2000, Hitachi Co., Ltd, Tokyo, Japan) operated at 200kV. TEM samples were tinned by focused ion beam system (FIB; FB-2100, Hitachi High-Technologies Co., Ltd, Tokyo, Japan).

RESULTS AND DISCUSSION

Densification behaviors

Figs. 2a and 2b show the changes in relative density (ρ) and temperature (T) as functions of time (t) for TZP[5] and TZP-B powder compacts, respectively. In the both kinds of compacts, the initial heating rate (H_R: slope of the T-t curve), the finally achieved temperature (T_{max}), the initial densification rate ($\dot{\rho}$: slope of the ρ-t curve) and the finally achieved density (ρ_{max}) increase with the increasing current. However, the period of time from start to finish of rapid densification for TZP-B is considerably shorter than that for TZP.

Fig. 3 shows the plot of $\dot{\rho}$ as functions of T at H_R=0.3, 1 and 2 K/s for TZP[5] and H_R=1, 2 and 3 K/s for TZP-B. T and $\dot{\rho}$ of the data point in Fig. 3 correspond to T at the intersection of the T-t curve and constant H_R curve (broken line) and the slope of the ρ-t curve at the intersection of the ρ-t curve and constant H_R curve (broken line), respectively (see Figs. 2a and 2b). Fig. 4 shows the plot of $\dot{\rho}$ as functions of ρ at H_R=0.3, 1 and 2 K/s for TZP[5] and H_R=1, 2 and 3 K/s for TZP-B. Similarly, ρ of the data point in Fig. 4 corresponds to ρ at the interaction of the ρ-t curve and constant H_R curve (broken line) in Figs. 2a and 2b. In Figs. 3 and 4, the $\dot{\rho}$-T and $\dot{\rho}$-ρ interpolation curves at fixed H_R are drawn not to go through on the other data points at various H_R (does not be displayed in Figs. 3 and 4) for accuracy. As shown in Fig. 4, although $\dot{\rho}$ for TZP-B is considerably larger than for TZP, there is no apparent difference in the $\dot{\rho}$-ρ line profiles between TZP and TZP-B. On the other hand, as shown in Fig. 3, T at the maximum $\dot{\rho}$ ($\dot{\rho}_{max}$) of the $\dot{\rho}$-T interpolation curves for TZP-B are considerably

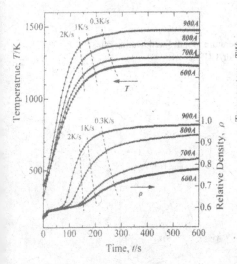
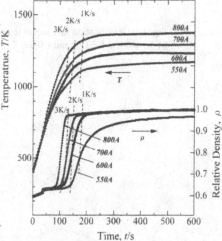

Fig. 2a Changes in relative density and temperature as functions of time for TZP consolidated at a fixed current of 600-900 A and applied stress of 46.5 MPa. "TZP" stands for ZrO_2-3 mol% Y_2O_3.

Fig. 2b Changes in relative density and temperature as functions of time for TZP-B consolidated at a fixed current of 550-800 A and applied stress of 46.5 MPa. "TZP-B" stands for (ZrO_2-3 mol% Y_2O_3)-20 mol% B.

lower than those for TZP. In addition, T at $\dot{\rho}_{max}$ for TZP-B (about 1120-1130 K) seems to be close to the onset temperature of rapid densification of TZP. These results clearly indicate the addition of boron leads to high-speed densification at lower temperatures.

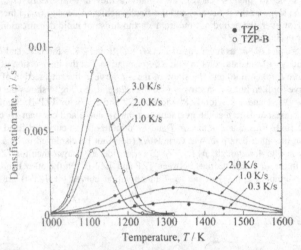

Fig. 3　Densification rate versus temperature plot at fixed heating rates of 0.3-2.0 K/s for TZP and 1.0-3.0 K/s for TZP-B.

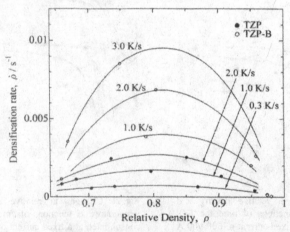

Fig. 4　Densification rate versus relative density plot at fixed heating rates of 0.3-2.0 K/s for TZP and 1.0-3.0 K/s for TZP-B.

Microstructure

It is difficult to suppress the crystallization and fabricate the amorphous TZP bulk samples, in spite of boron addition. Densification occurs after the fully crystallization of zirconia solid solution. Fig. 5 shows the typical XRD patterns for the TZP compacts consolidated at 600 to 900 A. There is no apparent difference in the three XRD patterns, and all the crystalline peaks correspond to the tetragonal zirconia solid solution (t-ZrO_{2ss}). On the other hand, as shown in Fig. 6, the strong crystalline peaks in the XRD patterns for the TZP-B compacts consolidated at 550 to 800 A are the monoclinic zirconia solid solution (m-ZrO_{2ss}). In addition, the weak crystalline peaks of YBO_3 and t-ZrO_{2ss} are also observed in Fig. 6. Although the ρ-t curves during cooling have not been presented in this paper, the volume expansion at around 1000 K was recognized for TZP-B but not TZP, suggesting the phase transformation from t-ZrO_{2ss} to m-ZrO_{2ss} during cooling after the finish of densification in the TZP-B samples.[7] Fig. 7 shows the SEM images obtained from the TZP and TZP-B samples consolidated at 900 and 800 A, respectively. Both samples show the similar equiaxed fine grain structure, but the average grain diameter of TZP-B (Fig. 7(b)) is considerably smaller than that of TZP (Fig. 7(a)). Although the presence of the minor phase of YBO_3 is recognized in the TZP-B samples by XRD (Fig. 6), the apparent minor phase area enough for measuring its composition, volume fraction and dispersion morphology could not be recognized in the SEM image.

Formation of m-ZrO_{2ss} in the TZP-B samples indicates the deficiency of Y_2O_3 (stabilizer of tetragonal structure of ZrO_{2ss}) in ZrO_{2ss} by addition of boron. It was reported[8] that m-ZrO_{2ss} and YBO_3 was formed by pressureless sintering of a mixed powder compact of 3Y-TZP and B_2O_3 by the reaction of

$$Y_2O_3 + B_2O_3 \rightarrow 2YBO_3 \tag{1}$$

However, the initial TZP-B powder before mechanical milling in this study does not contain B_2O_3 but B, and the B content (20.0 mol%) is considerably larger than the Y_2O_3 content (2.4 mol%), suggesting the remains of Y and B atoms after the YBO_3 formation. In order to discuss the structures and compositions of the minor phases, the Y_2O_3 content remained in m-ZrO_{2ss} (or the Y_2O_3 content ejected from ZrO_{2ss}) has to be estimated. Fig. 8 shows the relationship between the volume per unit cell and the Y_2O_3 content in m-ZrO_{2ss}. The straight line in Fig. 8 was drawn using the lattice parameters of pure m-ZrO_2 and m-ZrO_{2ss} containing 1.52 mol% Y_2O_3 (3 mol% $YO_{1.5}$) reported by Scott.[9] As shown in Fig. 8, the Y_2O_3 contents in the TZP-B compacts estimated by fitting the volumes per unit cell obtained by XRD (Fig. 6) to the straight line slightly decrease with increasing the current. However, the differences among them are considerably small, and the average value (for the samples consolidated at 600-800 A) of 1.54 mol% Y_2O_3 is quite similar to the solid solubility limit of Y_2O_3 in m-ZrO_{2ss}.[9] Assuming the solid solution of B in m-ZrO_{2ss} is negligibly small, the estimated composition of matrix / other minor phases is (77.60 mol% ZrO_2 + 1.23 mol% Y_2O_3) / (1.17 mol% Y_2O_3 + 20.00 mol% B). Thus the present YBO_3 phase is considered to be formed by the following reaction:

$$(ZrO_2)_{77.60}(Y_2O_3)_{2.40}(B)_{20.00} \text{ [amorphous]} \rightarrow$$
$$(ZrO_2)_{77.60}(Y_2O_3)_{1.23} [ZrO_{2ss}] + (YBO_3)_{1.17} + Y_{1.17}B_{18.83} \text{ [amorphous]} \tag{2}$$

Recently, it was suggested that the mixed powder compact of B and Y_2O_3 with a composition of B-5.53 mol% Y_2O_3, which is the same composition of the present minor phases of (1.17 mol% Y_2O_3 + 20.00 mol% B) mentioned above, consolidated by ECAS consists of YBO_3 together with YB_{12} and amorphous B.[7] The corresponding reaction is considered to be as follow:

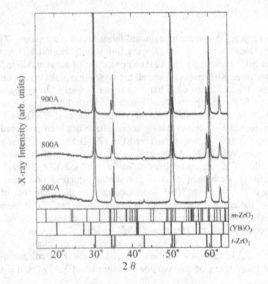

Fig. 5   X-ray diffraction patterns taken from TZP consolidated at 600, 800 and 900 A.

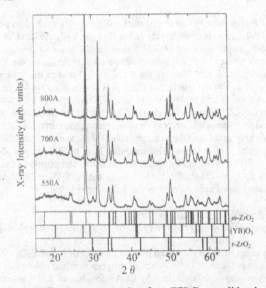

Fig. 6   X-ray diffraction patterns taken from TZP-B consolidated at 550, 700 and 800 A.

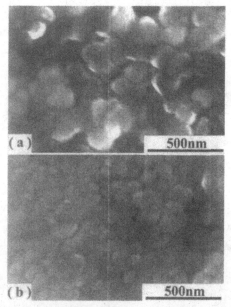

Fig. 7　SEM images of TZP consolidated at 900 A (a) and TZP-B consolidated at 800 A (b).

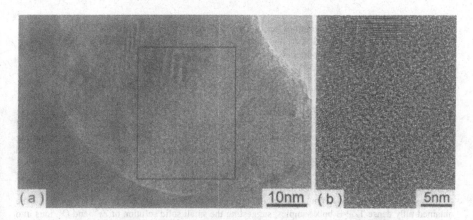

Fig. 8　TEM micrograph (a) of TZP-B consolidated at 800 A and the magnified micrograph (b) of the area enclosed with a black square in (a).

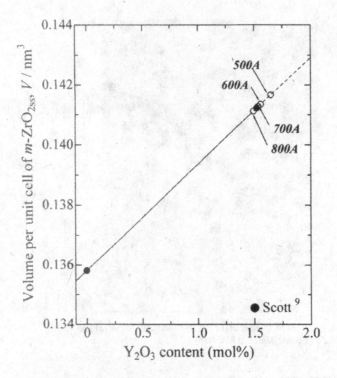

Fig. 9   Fitting of the volumes per unit cell of m-ZrO$_{2ss}$ obtained from TZP-B consolidated at 500-800 A to the solid line (volume per unit cell of m-ZrO$_{2ss}$ versus Y$_2$O$_3$ content) reported by Scott.[9]

$$(Y_2O_3)_{1.17} + (B)_{20.00} \rightarrow (YBO_3)_{1.17} + (YB_{12})_{1.17} + (B)_{5.96} \text{ [amorphous]} \qquad (3)$$

However, the crystalline peaks of yttrium borides,[10] namely YB$_{12}$, YB$_2$, YB$_4$, YB$_6$ and YB$_{66}$ could not be observed in the present study (see Fig. 6). These results suggest that the solid solution of Zr^{4+} and O^{2-} ions from ZrO$_{2ss}$ into B-rich amorphous (i.e., Y$_{1.17}$B$_{18.83}$ in Eq. (2)) occurs to some extent, resulting the prevention of the YB$_{12}$ formation. Indeed, the presence of an amorphous phase could be recognized in the TEM micrograph as shown in Fig. 9, although its composition has not been clarified yet. When the solid solution of Zr^{4+} and O^{2-} ions into B-rich amorphous is considerably small and the density of B-rich B-Y-Zr-O amorphous is close to the mixture of YB$_{12}$ and beta-B, the density of the TZP-B bulk could be calculated to 5.31 g/cm^3. This value is close to the measured density of 5.37 g/cm^3 for the obtained fully dense TZP-B bulk samples, suggesting the small solid solution of Zr^{4+} and O^{2-} ions into the B-rich B-Y-Zr-O amorphous. In addition, the calculated volume fractions of YBO$_3$ and B-Y-Zr-O amorphous are 1.6 and 14.6 vol. %, respectively, indicating the minor phases consist mainly of the B-Y-Zr-O glassy phase.

Stress exponent and activation energy

Densification behaviors can be expressed by the following equation[11]:

$$\dot{\rho}/\rho = A/T(b/d)^p(\sigma_{eff}/G_{eff})^n D_0 \cdot exp(-Q/RT) \tag{4}$$

where A is a dimensionless constant, T is the absolute temperature, b is the magnitude of Burgers vector, d is the grain size, p is the grain size exponent, σ_{eff} is the effective applied stress, G_{eff} is the effective shear modulus, n is the stress exponent, D_0 is the diffusion coefficient, Q is the activation energy and R is the gas constant. Since the X-ray peak-broadening at half-maximum of m-ZrO_{2ss} for TZP-B and t-ZrO_{2ss} for TZP corresponding to the grain size (i.e., the Scherrer method) is almost constant, independent of the densification temperature (see Figs. 5 and 6), Eq. (4) can be simplified to as follows:

$$\dot{\rho}/\rho = A'/T(\sigma_{eff}/G_{eff})^n D_0 \cdot exp(-Q/RT) \tag{5}$$

Assuming the negligibly small influence of the minor B-Y-Zr-O and YBO_3 phases on the bulk shear modulus, the effective shear modulus of the porous TZP and TZP-B samples can be expressed by the following equation[12]:

$$G'_{eff}[MPa] = 66.051\times10^3 - 116.003\times10^3 \times(1-\rho) \tag{6}$$

The temperature-compensated shear modulus of the porous TZP and TZP-B samples can be drawn as follows[13]:

$$G_{eff}[MPa] = G'_{eff}\times10^3 - 13.3T \tag{7}$$

In addition, σ_{eff} is expressed by using ρ and initial relative density (ρ_0), [14] as

$$\sigma_{eff}/\sigma = (1-\rho_0)/\rho^2(\rho-\rho_0) \tag{8}$$

Fig. 10 shows the relationship between σ_{eff} compensated for the threshold stress (σ_0) (i.e., (σ_{eff} - σ_0) instead of σ_{eff}) and $\dot{\rho}$ for TZP[5] in the temperature range from 1150 to 1400 K. ρ is also shown at the side of each data point. All the least squares lines indicate the similar slope of n = 2, even near the boundary of the initial ($\rho<0.92$) and final ($\rho \geq 0.92$) stages. In the case of TZP-B (Fig. 11), on the other hand, the least squares lines in the low stress region indicate the similar slope of n =1.35 without compensation of σ_{eff} for σ_0. However, the n value changes from 1.35 to 2 with an increase of T (and/or a decrease of σ_{eff}). Fig. 12 shows Arrhenius plots of the (σ_{eff}- σ_0)-compensated $\dot{\rho}$ for TZP[5] and σ_{eff} -compensated $\dot{\rho}$ for TZP-B, respectively. All the data points for TZP are well on the one straight line, and the estimated Q value is 460 kJ/mol, which is quite similar to the value of 460±40 kJ/mol obtained by creep experiments.[3,4] On the other hand, one straight line could not be drawn for TZP-B, and the Q values estimated from the two best fitted least squares lines are 582 and 562 kJ/mol for the low and high temperature region, respectively. It should be noted that the transition temperature of Q for TZP-B (about 1120 K) corresponds to the onset temperature of rapid densification of TZP (see Fig. 4), and the transition temperature of n=1.35 to 2 for TZP-B (see Fig. 11).

These results clearly indicate that the additive of boron strongly affects the densification behavior of TZP during EACS. Based on the estimated n and Q values in Figs. 10, 11 and 12, the densification process is considered to be controlled by the single mechanism for TZP (i.e., grain boundary sliding), but not for TZP-B. In the case of TZP-B, in the low temperature region (T: 1120 K), the accommodation of the stress concentration generated by sliding and rotation of the matrix ZrO_{2ss} grains seems to be occurred by the dissolution of Zr and O ions into the minor phase (mainly B-rich B-Y-Zr-O amorphous phase), since the ZrO_{2ss} grains could not be able to deform plastically without the help of the minor phase. The small value of $n=1.35$ may suggest the relatively insensitive property of the dissolution-precipitation reaction through the ZrO_{2ss}/B-Y-Zr-O interface (and/or diffusion of Zr and O ions through the B-Y-Zr-O phase) for the stress concentration morphology in the matrix ZrO_{2ss} grains. In the high temperature region ($T>1120$ K), since the matrix ZrO_{2ss} grains can deform without the help of the other phase, the accommodation of the stress concentration in the matrix ZrO_{2ss} grains could be occurred by diffusion along the matrix/amorphous (and/or matrix/matrix) interfaces, resulting the similar n value of TZP.

Since the physical properties of the B-rich B-Y-Zr-O glassy phase (i.e., solid solubility of ZrO_{2ss}, viscosity, shear modulus) have not been clarified, we could not obtain the exact values of n and Q for TZP-B by using Eq. (5). In addition, since the distribution morphology of the B-rich B-Y-Zr-O glassy phase has also been unknown, we could not deny the possibility of densification only by viscous flow of the glassy phase in the low temperature region mentioned above (i.e., no occurrence of stress concentration if the glassy phase always exists between the matrix ZrO_{2ss} grains). Further studies of the B-rich B-Y-Zr-O glassy phase are needed for analyzing quantitatively the densification behaviors of TZP-B.

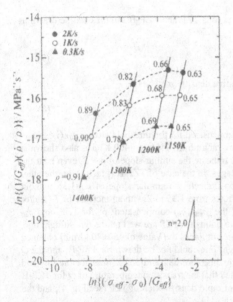

Fig. 10 Plot of effective stress (σ_{eff}) with compensation for threshold stress (σ_0) versus densification rate for TZP.

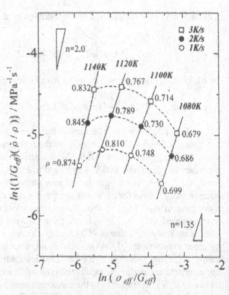

Fig. 11 Plot of effective stress (σ_{eff}) without compensation for threshold stress (σ_0) versus densification rate for TZP-B.

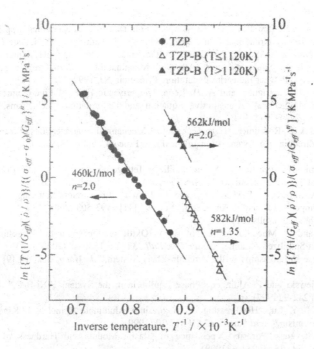

Fig. 12 Arrhenius plots of densification rate compensated for temperature and threshold stress (σ_0) for TZP and for temperature only for TZP-B.

CONCLUSION

The rapid densification behaviors of the mechanically milled TZP and TZP-20 mol% B powders have been investigated. The addition of boron results in the higher-speed densification of TZP at lower temperatures. Although the TZP bulk consists of only the tetragonal zirconia solid solution (t-ZrO_{2ss}), the TZP-B bulk consists of the matrix phase of monoclinic zirconia solid solution (m-ZrO_{2ss}) and the minor phases of B-rich B-Y-Zr-O glassy phase and YBO_3. The estimated n and Q values are constant for TZP ($n=2$ and $Q=460$ kJ/mol), but changes from the low temperature region ($n=1.35$ and $Q=582$ kJ/mol) to the high temperature region ($n=2$ and $Q=562$ kJ/mol) for TZP-B. Since the transition temperature of about 1120 K for TZP-B has good agreement with the onset temperature of rapid densification of TZP, densification of TZP-B seems to occur by the dissolution-precipitation at the matrix / minor glassy phase interfaces (and/or diffusion through the minor glassy phase) at the low temperature region and grain boundary sliding of the matrix phase at the high temperature region.

REFERENCES

[1] R. Orrú, R. Licheri, A. M. Locci, A. Cincotti and G. Cao, "Consolidation/Synthesis of Materials by Electric Current Activated/Assisted Sintering," *Mater. Sci. Eng. R*, **63** [4-6] 127-287 (2009).

2   H. Kimura, "Non-Equilibrium Powder Processing of Full Density Nanoceramics," pp.55-61 in Proceedings of the 1999 International Conference on Powder Metallurgy & Particulate Materials (PM^{2}TEC'99), Advances in Powder Metallurgy & Particulate Materials, Part12, Intermetallics, Nanophase Materials & Shape Memory Alloys / Machinability / Structure, Properties & Characterization, Metal Powder Industries Federation, Princeton, NJ, 1999.

3   J. M. Melendo, A. D. Rodríguez, and A. B. León, "Superplastic Flow of Fine-Grained Yttria-Stabilized Zirconia Polucrystals: Constitutive Equation and Deformation Mechanisms," *J. Am. Ceram. Soc.*, **81** [11] 2761-76 (1998).

4   M. J. Melendo and A. D. Rodríguez, "High Temperature Mechanical Characteristics of Superplastic Yttria-Stabilized Zirconia. An Examination of the Flow Process," *Acta Mater.*, **48** [12] 3201-10 (2000).

5   T. Kumagai, "Rapid Densification of Yttria-Stabilized Tetragonal Zirconia by Electric Current-Activated/ Assisted Sintering Technique," *J. Am. Ceram. Soc.*, to be published (2011).

6   M. Dourandish, A. Simchi, E. T. Shabestary, and T. Hartwig, "Pressureless Sintering of 3Y-TZP/ Stainless-Steel Composite Layers," J. Am. Ceram. Soc., **91** [11] 3493-503 (2008).

7   T. Kumagai, unpublished results (2010).

8   D. Z. de Florio and R. Muccillo, "Effect of Boron Oxide on the Cubic-to-Monoclinic Phase Transition in Yttria-Stabilized Zirconia," *Mater. Res. Bull.*, **39** [10] 1539-48 (2004).

9   H. G. Scott, "Phase Relationships in the Zirconia-Yttria System," *J. Mater. Sci.*, **10** [9] 1527-35 (1975).

10 H. Szillat, P. Majewski and F. Aldinger, "Phase Equilibria in the System Y-Ni-B-C," *J. Alloys. Compd.*, 261 [1-2] 242-9 (1997).

11 C. J. Ting and H. Y. Lu, "Hot-Pressing of Magnesium Aluminate Spinel – I. Kinetics and Densification Mechanism," *Acta Mater.*, **47** [3] 817-30 (1999).

12 J. Luo and R. Stevens, "Porosity-Dependence of Elastic Moduli and Hardness of 3Y-TZP Ceramics," *Ceram. Int.*, **25** [3] 281-6 (1999).

13 N. Balasubramanian and T. G. Langdon, "Comment on the Role of Intragranular Dislocations in Superplastic Yttria-Stabilized Zirconia," *Scripta Mater.*, **48** [5] 599-604 (2003).

14 E. Arzt, M. F. Ashby, and K. E. Easterling, "Practical Applications of Hot-Isostatic Pressing Diagram: Four Case Studies," *Metall. Trans. A*, **14A** [2] 211-21 (1983).

MICROSTRUCTURE CONTROL OF Si_3N_4 CERAMICS USING NANOCOMPOSITE PARTICLES PREPARED BY DRY MECHANICAL TREATMENT

Junichi Tatami[1], Makoto Noguchi[1], Hiromi Nakano[2], Toru Wakihara[1], Katsutoshi Komeya[1] and Takeshi Meguro[1]
1 Yokohama National University, Yokohama, Japan
2 Toyohashi University of Technology, Toyohashi, Japan

ABSTRACT
    Microstructure control is essential for improving the properties of fabricated ceramics. In this study, the microstructure of Si_3N_4 ceramics was controlled using nanocomposite particles consisting of submicron Si_3N_4 particles with Y_2O_3 and Al_2O_3 nanoparticles as sintering aids. These nanocomposite particles were prepared by dry mechanical treatment. The prepared nanocomposite particles were molded by uniaxial pressing and cold isostatic pressing. The green body thus obtained was fired at 1800 °C for 2 h in 0.9 MPa N_2. For comparison, a powder mixture prepared by conventional wet ball milling was used to obtain another sintered body. The specific surface area of the powder mixture prepared by dry mechanical treatment decreased with an increase in the applied power and treating time; this implied that the nanocomposite particles were formed by the bonding of nanoparticles with a submicron particle. β-Si_3N_4 grains in the Si_3N_4 ceramics fabricated using the nanocomposite particles prepared by dry mechanical treatment were more elongated than those in the Si_3N_4 ceramics fabricated using powder mixture prepared by wet ball milling. The Si_3N_4 powder prepared by dry mechanical treatment showed higher fracture toughness because of its more elongated β-Si_3N_4 grains.

INTRODUCTION

    Si_3N_4 ceramics were first fabricated more than 40 years ago. Over the course of time, the fabrication of Si_3N_4 ceramics with high strength and fracture toughness was made possible by the development of the following: SiAlONs [1,2]; sintering aids such as Y_2O_3 [3,4]; fine, pure, and highly sinterable Si_3N_4 powder [5]; a gas pressure sintering technique [6]; advanced science and technology for microstructure control; and so on. Si_3N_4 ceramics have been applied to automobile components such as glow plugs [7], hot chambers [8], and turbocharger rotors [9]. During the same period, cutting tools and bearing components were also developed [10,11]. Although the cost of Si_3N_4 ceramics was high, they were used as bearing materials in machine tools owing to their advantageous properties such as high strength, toughness, elastic modulus, and hardness, as well as light weight and good corrosion resistance.

    Si_3N_4 ceramics with elongated β-Si_3N_4 grains exhibit excellent mechanical properties. The use of nanoparticle dispersion is convenient and effective in controlling a microstructure on a nanoscale, because nanoparticles lead to an improvement in sinterability and reactivity, unlike submicron particles. Furthermore, the homogeneous mixing of additives can be better achieved using nanoparticles rather than large particles as a raw material. However, the use of wet ball milling with liquid dispersion media such as water and ethanol might lead to the reagglomeration of nanoparticles during drying. We focused on the fabrication of advanced ceramics by using a powder composite process in which nanoparticles are mechanically mixed using only a dry mixing process that enables the fabrication of nanocomposite particles. In this process, nanoparticles bind to submicron particles because of the application of an external mechanical force - specifically, a shear force [12].

    Some researchers have synthesized ceramic particles and prepared porous and dense ceramics by a dry mechanical treatment [13-18]. In this paper, we present the microstructure and properties of Si_3N_4 ceramics fabricated using nanocomposite particles prepared by a dry mechanical treatment process.

EXPERIMENTAL PROCEDURE

High-purity, fine Si$_3$N$_4$ powder (SN-E-10, Ube Industries Ltd., Japan) was used as a raw material. Y$_2$O$_3$ (BB, Shin-etsu Chemical Co., Ltd., Japan) and γ-Al$_2$O$_3$ (TM-300, Taimei Chemical Co., Japan) were added to the Si$_3$N$_4$ powder as sintering aids. The batch composition was in the ratio Si$_3$N$_4$:Y$_2$O$_3$:Al$_2$O$_3$ = 92:5:3 (wt%). Nanocomposite particles were prepared using a powder composer (Nobilta NOB-130, Hosokawa Micron Co., Japan). First, γ-Al$_2$O$_3$ and Si$_3$N$_4$ were put into the powder composer to be pre-mixed at 200 rpm for 5 min. After premixing, the powder mixture was mechanically treated at 5 kW for 10 min. After that, Y$_2$O$_3$ nanoparticles were also added to the powder mixture of γ-Al$_2$O$_3$ and Si$_3$N$_4$ followed by the mechanical treatment again at the same condition. 4 wt% paraffin (melting point: 46–48 °C, Junsei Chemical Co., Japan) and 2 wt% dioctyl phthalate (DOP, Wako Junyaku Co., Japan) were added as a binder and lubricant, respectively. The mixed powders were sieved using a #60 nylon sieve; then, they were molded into φ 15 × 7 mm pellets by uniaxial pressing at 50 MPa and subsequent cold isostatic pressing at 200 MPa. After binder burnout in air at 500 °C for 3 h, the obtained green bodies were fired at 1800 °C for 2 h in 0.9 MPa N$_2$ using a gas pressure sintering furnace (Himulti 5000, Fujidenpa Kogyo Co., Japan). The density of the fired samples was measured by the Archimedes method. The phases present in the samples were identified by X-ray diffraction (RINT2000, Rigaku Co., Japan).

RESULTS AND DISCUSSION

Figure 1 shows SEM photographs of the powder mixture prepared by mechanical treatment and wet ball milling. Although many nanoparticle agglomerates were found in the powder mixture

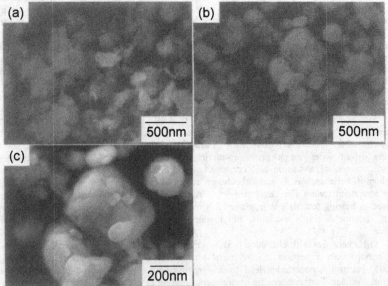

Fig.1 SEM photographs of the powder mixtures prepared by (a) wet ball milling and (b) mechanical treatment. (c) is an enlarged view of (b).

after wet ball milling, nanosized γ-Al$_2$O$_3$ and Y$_2$O$_3$ particles were found to be uniformly dispersed on the submicron Si$_3$N$_4$ particles. Fig. 2 shows the specific surface area of the powder mixture prepared by mechanical treatment. After the mechanical treatment of Si$_3$N$_4$ and γ-Al$_2$O$_3$, the specific surface area of the powder mixture decreased with an increase in the applied power. This probably resulted from the bonding of γ-Al$_2$O$_3$ nanoparticles with a Si$_3$N$_4$ particle, similar to that observed in the case of the mechanical treatment of γ-Al$_2$O$_3$ nanoparticles with ZnO submicron particles. The specific surface area of the powder mixture after the addition of Y$_2$O$_3$ and mechanical treatment also decreased slightly with an increase in the applied power.

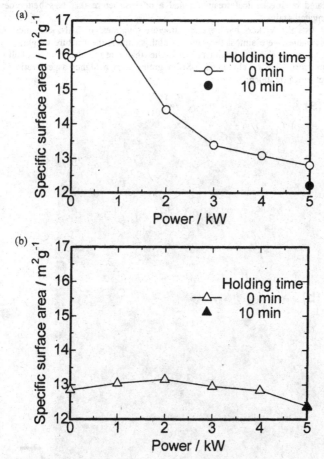

Fig. 2 Specific surface area of the nanocomposite particle of (a) Si$_3$N$_4$-Al$_2$O$_3$ and (b) Si$_3$N$_4$-Al$_2$O$_3$-Y$_2$O$_3$.

Fig. 3 shows the SEM photograph of the plasma etched surfaces of the Si$_3$N$_4$ ceramics prepared by wet ball milling and mechanical treatment. Elongated β-Si$_3$N$_4$ grains were formed in the ceramics prepared by both wet ball milling and mechanical treatment. Fig. 4 shows the grain-size distribution (the length of the short axis) evaluated by an image analysis of the SEM photographs. The grain size distribution broadened slightly in the Si$_3$N$_4$ ceramics fabricated by mechanical treatment; however, it remained almost the same as that in the Si$_3$N$_4$ ceramics fabricated by ball milling. The average lengths of the short axis after wet ball milling and mechanical treatment were 0.26 and 0.24 μm, respectively. However, the aspect ratios were 3.8 and 4.5, respectively; this implies that the Si$_3$N$_4$ ceramics fabricated by mechanical treatment had a microstructure that was better developed and consisted of elongated grains.

Table 1 lists the Vickers hardness and fracture toughness of Si$_3$N$_4$ ceramics. Although the Vickers hardness values were almost the same in both samples, the fracture toughness of the Si$_3$N$_4$ ceramics fabricated by mechanical treatment was higher than that fabricated by wet ball milling. This resulted from the formation of elongated β-Si$_3$N$_4$ grains with a higher aspect ratio in the sample obtained by mechanical treatment.

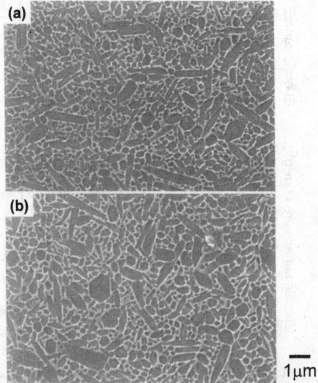

(a)

(b)

1μm

Fig.3 SEM photographs of plasma etched surfaces of Si3N4 ceramics prepared by (a) wet ball milling and (b) mechanical treatment.

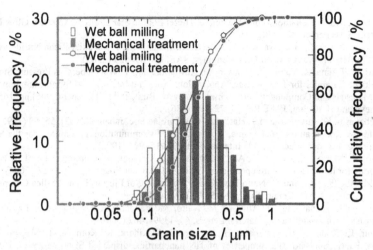

Fig. 4 Grain size distribution of the Si₃N₄ ceramics.

Table 1 Mechanical properties of the Si₃N₄ ceramics

	Wet ball milling	Mechanical treatment
Vickers hardness, H_V (GPa)	14.2	14.0
Fracture toughness, K_{IC} (MPa m-1/2)	7.0	7.5

CONCLUSIONS

Nanocomposite particles consisting of Si₃N₄–nano-Al₂O₃–nano-Y₂O₃ were prepared by dry mechanical treatment to fabricate Si₃N₄ ceramics. Elongated β-Si₃N₄ grains with a high aspect ratio were formed in the Si₃N₄ ceramics fabricated using the nanocomposite particles prepared by mechanical treatment. As a result, the fracture toughness of these Si₃N₄ ceramics was higher than that of the Si₃N₄ ceramics fabricated by wet ball milling.

REFERENCES

[1]Y. Oyama and S. Kamigaito, Solid Solubility of Some Oxides in Si₃N₄, Jpn. J. Appl. Phys., 10[11], 1637 (1971).
[2]K. H. Jack, Sialons and related nitrogen ceramics, J. Mater. Sci., 11, 1135-1158 (1976).
[3]A. Tsuge, K. Nishida and M. Komatsu, Effect of Crystallizing the Grain-Boundary Glass Phase on the High-Temperature Strength of Hot-Pressed Si₃N₄ Containing Y₂O₃, J. Am. Ceram. Soc., 58[7–8], 323–326 (1975).
[4]K. Komeya, Am. Ceram. Soc. Bull., 63[9], 1158–1159, 1164 (1984).
[5]T. Yamada and Y. Kotoku, Jpn. Chem. Ind. Assoc. Mon., 42, 8 (1989).
[6]M. Mitomo, J. Mater. Sci., Pressure sintering of Si₃N₄, 11[6], 1103–1107 (1976).

[7]H. Kawamura and S. Yamamoto, Improvement of Diesel Engine Startability by Ceramic Glow Plug Start System, SAE Paprer, No. 830580 (1983).

[8]S. Kamiya, M. Murachi, H. Kawamoto, S. Kato, S. Kawakami and Y. Suzuki, Silicon Nitride Swirl Chambers for High Power Charged Diesel Engines, SAE, No. 850523 (1985).

[9]H. Hattori, Y. Tajima, K. Yabuta, Y. Matsuo, M. Kawamura and T. Watanabe, Gas Pressure Sintered Silicon Nitride Ceramics for Turbocharger Application, Proc. 2nd International Symposium bon Ceramic Materials and Components for Engines, Ed. by W. Bunk and H. Hausner (1986) pp. 165.

[10]K. Komeya and H. Kotani, JSAE Rev., 7, 72–79 (1986).

[11]H. Takebayashi, K. Tanimoto and T. Hattori, J. Gas Turbine Soc. Japan, 26[102], 55–60 (1998).

[12]M. Naito, A. Kondo and T. Yokoyama, Applications of Comminution Techniques for the Surface Modification of Powder Materials, ISIJ International, 33, 915–924 (1993).

[13]T. Fukui, K. Murata, S. Ohara, H. Abe, M. Naito and K. Nogi, Morphology control of Ni–YSZ cermet anode for lower temperature operation of SOFCs, J. Powder Sources, 125, 17-21 (2004).

[14]H. Abe, I. Abe, K. Sato and M. Naito Dry Powder Processing of Fibrous Fumed Silica Compacts for Thermal Insulation, J. Am. Ceram. Soc., 88, 1359-1361 (2005).

[15]K. Sato, J. Chaichanawong, H. Abe and M. Naito, Mechanochemical synthesis of LaMnO$_3$ fine powder assisted with water vapor, Material Letters, 60, 1399-1402 (2006).

[16]J. Tatami, E. Kodama, H. Watanabe, H. Nakano, T. Wakihara, K. Komeya, T. Meguro and A. Azushima, Fabrication and wear properties of TiN nanoparticle-dispersed Si3N4 ceramics, J. Ceram. Soc. Japan, 116, 749-754 (2008).

[17]S. Tasaki, J. Tatami, H. Nakano, T. Wakihara, K. Komeya and T. Meguro Fabrication of ZnO ceramics using ZnO/Al$_2$O$_3$ nanocomposite particles prepared by mechanical treatment, J. Ceram. Soc. Japan, 118, 118-121 (2010).

[18]D. Hiratsuka, T. Junichi, T. Wakihara, K. Katsutoshi and T. Meguro Fabrication of AlN ceramics using AlN and nano-Y$_2$O$_3$ composite particles prepared by mechanical treatment, Key Engineering Materials, 403, 245-248 (2009).

Author Index

Author Index

Printed in the United States
By Bookmasters